Table of Contents

Table of Contents	2
Introduction	3
The Bending of Mathematics	6
Holes	10
Special Relativity	17
Gravity as the Odd Ball	23
Space-Time Curvatures	26
4-D Curvature	30
The Sun's Magnetic Poles	33
The Bumper Effect	36
Solar Collectors and Solar Cells	41
The Evolution of Batteries	45
Endothermic Chip	49
Propellent-less Rocket	56
$E=mc^2$	63
Relativistic Mass	66
Quantum Entanglement	68
Quantum Telescope	71
Quantum Communication	74
Quantum Teleportation	77
Mitochondria	79
Venus	83
The Birth and Composition of Neutron Stars and Blackholes	85
The Edge of the Universe	88
The Economy of Space Mining	91
Western US Drought	95
Other Considerations for the Spaceflight to Mars	106
The Earth does not Orbit the Sun	111
Nuclear Physics	114
Friction	118
The Sound of the Sun	121
The Not So Standard of Mass	123
Applications of Science	125
Credits	128

Introduction:

This book is going to cover topics in math and physics. Despite covering math topics, I am trying to keep the equations to a minimum as I am more geared towards the non-physicist more so than the physicist. Physics is often based on models and interpretation. In interpretation, we find that different interpretations can exist, such as the Standard Model and String Theory, yet they do not seem to fit with one another yet. For each theory, they are valuable in their own right and offer insight into physics, despite the fact that they do not work with one another. To that end, I shall offer my own interpretation on topics. Many people are quick to simply jump in and say a topic I cover and offer an interpretation is just wrong. Some of the greatest professors that I have worked with instead say that this is not my strong suit.

I must say that like many physicists, I run into an issue of miscommunication of being unclear and misunderstood for several reasons. First, I have run ideas through my head for decades at times to make sense of a topic and sometimes, I skip over some of the steps that I have made to make a concept work. I will also be the first to admit here that as in any time someone comes up with a concept, until proven one way or another, some might be right and some could be completely wrong. If a concept is wrong and I am using a wrong premise, building upon a mistaken concept will lead to further errors down range. At the risk of being wrong, I am still offering bold ideas at the risk of possibly being embarrassed. Despite that, it is still worth the risk in case I might be right one even one or two big concepts. Second, I am not an English major nor public speaker and offering a clarity in an English format, ensuring I offer a premise first, a body, and a conclusion, along with any other nuances that are necessary, this will often be a challenge for me. This is also the first edition, I anticipate mistakes. My old high school English teacher always told me not to write in the passive, but active tense to keep things interesting. Sadly, I will fail at this many time over as well, though I do not really see how that will change some of the profound concepts that I offer here. Third, physics is not just a language best spoken in mathematics, but also a bending of math that often a mathematician does not speak. I am making this concept alone the first topic in this book.

I usually like taking notes within the books that I read, but margins are not usually enough. For that reason, after each topic, I am leaving a blank page or two depending on the topic for you to take notes or questions or doodling, whatever you like.

This book is not only for looking at a new technology and theories. I am also pointing out many existing technologies, some of which I believe are underutilized. I have attempted to make it clear which are established and which may be novel. Some even lead to simple questions that I do not have an answer for. Sometimes those are the most profound questions that push for further exploration.

The Bending of Mathematics

The bending of math is largely one of the greatest differences between mathematics and physics. As a student, I had a friend that was all in for being a physics major. Into the second year, she quickly realized that she did not have the ability to bend math as would be needed. She changed her major to mathematics quickly. To be honest, her math skills are a hundred times better than mine.

As an example, if we are asked to add three vectors, all moving at different angles, a mathematician will create one equation with the three vectors as presented and crunch the numbers from there. Physicists are a bit different in that we will just stare at the vectors for a moment and find a convenient way of rotating the $\hat{x}$ and $\hat{y}$ axis such that the new $\hat{x}$ axis lies on one of the three vectors. This simplifies the resulting equation into 2 vectors to write an equation for and crunch numbers for rather than 3 vectors.

In main stream mathematics, it is often easy to get a derivative of an equation and see that if a variable x represents a length, the derivative being $\frac{dx}{dt}$ is representing a change in length with respect to time, yielding an equation for velocity. There is nothing really novel here. However, if we consider a derivative of $\frac{dx}{dy} = (x+2)$ as a fairly easy equation, it is not uncommon to separate the derivative and work from there, yielding $dx = (x+2) \cdot dy$. Sometimes crunching harder equations requires this splitting is required though to be honest, this split equation is not something that is making sense and will eventually have to be reformulated in a form that can be visualized. However, there is another factor that allows us to use this process to be useful that did not exist in the past. With modern technology, we can do something like use a device to measure radiation coming off of a material, offer a million data points or more, and then good math programs will make a single equation that fits them really nicely. When I say nicely, that may be a bit of a misnomer though. The equation might be several pages long and not for people to visualize, but we can ask the computer to take a derivative of this page long equation and it can do it within seconds. It can even give us information like its deviations and such. The computer can even give us a graph so that we can visualize something that we cannot see in the raw page's long equation. It is really something that we have many tools at our disposal that over the last hundred years or even the last twenty years, we have not had. It is a modern marvel. I think this is a good starting point to lead us into our next topic of holes in equations.

Holes

In physics, I find that there are basically two types of physicists, regardless of their sub-field or title (such as department head that focuses on administration heavily). I see that theoretical physicists make up about 10% of physicists while the remaining 90% are experimental physicists. It is not a degree or specialized classes that separates the class and one is certainly not better than the other. It is the individual scientist that finds themselves in that category. Each heavily rely on the other to create a work of art.

To that end, let us start with the most basic equation that everyone can understand from basic algebra.

$$y = (x+1)$$

This is a straight line with a slope. Something we all know is that we should be able to multiply an equation by the value of one and get the same result. However, there is an exception to this rule. What might complicate this might be where the computer creates that pages long equation for us and we might want to jump in and simplify the equation a touch. So, let us multiply by a factor of one, being $\frac{(x-2)}{(x-2)}$, which should be fine, but it is not really. These yields:

$$y = \frac{(x+1)(x-2)}{(x-2)}$$

We now have a problem with the equation. For almost all of the equation, it works out just fine. If we look at the points just around 2, we have a hole in the graph. For the numerator, it is easy to plug a value in and get a relatively small number.

However, any point just around 2 is going to give us a value of zero. This is a problem because a zero in the denominator yields an asymptotic value.

A mathematician might be quick to jump in and say the value is convergent according to math principles or maybe other reasons as can be proved. I would say sure, mathematically okay. This is also where physicists have to be able to look at pages long equations generated by a computer and see these holes that the computer would use principles saying, oh, they just cancel out. However, the physicist needs to find those canceling points and call them out as critical points.

Any time you get into a point where you might find a critical point, one being an asymptote, this is where math is really breaking down. A theorist needs to point this out and say that something funny is going on and it needs to be checked out.

In my first year of physics, we were always given a list of equations on tests or quizzes. We were not expected to memorize them. Even into the second year, we were given the same, however, we quickly learned our dividing point between physics and math. In the movie "Arrival," the linguist in the group described a phenomenon where as a person learns a different language, their brains are wired differently to think a different way because of the language. In the movie, she even demonstrates this where as she learns the language, she begins to experience a thought process in non-linear terms. To this end, she begins to see the future. While the movie is awesome in its own right, it is unfortunate that 99% of movie goers do not really understand non-linear and the movie did not explain it well. So, the movie goer might see that she is seeing the future, they don't understand the nuance of the non-linear thinking.

To that end, in the second year of physics we begin to really see that only a couple of the equations that we are given, only a couple really stick out as important. Out of all of the basic equations of Newton's motion, I picked out one for velocity as a starting point for example. I could have probably picked out another arbitrarily and would end up with the same results or at least similar enough, but let's make that a starting point and see where it takes us.

$$v = \frac{dx}{dt} + x$$

Very simply, as we look in the change in the x direction as time passes, we can find a position. I shall hit on the importance of *a starting point* momentarily. With this starting point, I no longer need to look at the list of equations that I have been given, but know how to quickly manipulate the equation for some factors that I am wanting. If I want to know an equation for acceleration without looking at the equation sheet, here is my thought process with a derivative, how is that velocity changing over time:

$$\frac{dv}{dt} = \frac{dx}{dt} \div dt = \frac{dx}{dt} \div dt = \frac{dx}{dt^2} = a + dx$$

With *a* representing acceleration and some fudge factor we can look at experimentally and see there was a starting point x in there. We can also integrate the original and take the area under the curve to find the distance we had traveled. This same thought process, which we quickly learn to manipulate in our heads, it can apply to many types of equations. Very similarly, we might be given several equations within a system and rather than try and manipulate with algebra and a system of equations, we might quickly recognize this as a matrix and solve it using that method.

Going back to our basic equation of a hole, as a mathematician might be able to use their methods to determine this anomaly as a certain value, a physicist must look at an infinity in an equation and view it as a critical point where the experimentalist physicist must carefully evaluate what is going on at that point. As an easy example, here is the equation for gravity:

$$F_g = \frac{Gm_1m_2}{r^2}$$

This equation appears to work fine for the moon or Earth, but the trained physicist knows the most important thing to do is to see an equation and look for where is the funny business going on at? These are our *critical points*. Here, we see that if somehow our value of r (radius) becomes zero, our idea of F_g is going to start to break down and not make much sense. Do we see this in practice? Indeed, we do. So much so that Einstein himself questioned and doubted his own work. He looked at the equations and imagined a blackhole (even before it was named) and said it could not exist in reality. But, we do in fact find instances of blackholes that do exist. The math begins to not work at that strange point where all of its matter (more than in the Sun) collapses to a radius of zero though. If in fact all of that mass collapsed to exactly zero or even just a hair bigger than we think it does, the math shows an asymptotic value such that the blackhole would have an event horizon that would encapsulate the Milky Way entirely. Yet, we don't find that to be the case. In reality, a blackhole can have an event horizon of different sizes based on the mass of the blackhole, but rather than strictly looking at the force of gravity, it does not work and we have created other equations to represent a blackhole instead. We can use a different equation such as a Schwarzschild radius.

and

You may come across variations of the equation that are equivalent, but there is an important *critical point* that we are looking at. If the velocity of a spacecraft or a probe manages to have a velocity of c, the speed of light, the equation comes across one of those points where we are dividing by zero. Very often this is interpreted as this is the proof that to get a probe moving the speed of light, we would need to have a fuel source that is infinite as well and therefore traveling the speed of light is impossible. I am going to challenge this interpretation for two main reasons.

First, as I have previously covered, this is a *critical point* such that when we are dealing with a point of an infinite, this is not the place to rush to a conclusion, rather one in which we find a point that needs to be handed over to an experimentalist for further evaluation.

Second, as we consider Cherenkov radiation, when a charged particle like an electron is passed through a dielectric medium at a greater speed than the phase velocity of a medium in that of light. This is similar to that of a sonic boom of sound, but instead, it is the sonic boom of light.

This not only demonstrates that passing the speed of light is possible, but further, we know that we can expect a version of this Cherenkov radiation in the process. To this end, not only do I propose that we ought to be handing the process to experimentalists to begin work, but also, we have a beginning point to tell us to expect a radiation to occur as a result. Also, we can probably tell that the Lorentz Transformation equations are probably going to break down and unlike what they suggest, we will not likely have a key into traveling into the past, but instead will find something else happen and whatever this thing is, it will probably involve radiation. This leads us straight into Space-Time Curvatures STC.

Special Relativity

I often forget that people struggle with the implications of special relativity. An understanding of this is going to be really important when the topic of space-time curves (STC) comes up.

Special relativity was built on a single premise. Einstein developed one theory that demanded that the speed of light be equal for all observers. From there, he built equations of velocity and time that would fit this requirement. We can consider a separate case in which two spaceships are travelling at 99% of the speed of light towards each other. Normally, math would sum the two values and say that there is going to be a resultant of 198% the speed of light collision. This turns out not to be the case though. Because of the initial demand that even if you are traveling at 99% of the speed of light, the only way that you could still see light from the other spaceship coming at you is that the time aboard your ship (with respect to an observer on earth watching this collision) must slow down a lot. With your spaceship moving quickly and time slowed, you observe the light wave's peaks coming in slower so nothing seems out of place for you now. Time and space are one thing though.

Here, t' represents the time aboard the spaceship moving fast, v represents the spaceship velocity, and c represents the speed of light. There is a list of a few principles that physicists are taught early on at the university. One of these is to take the equation and find where it seems to blow up. In this case, as v approaches the value of c, the denominator of the equation approaches zero. This makes the value of t' to approach infinity, implying that time is stopping. This is not just a play on the equations, but a real quality that is occurring quantitively.

You will recall from the previous topic of holes, that infinity may play itself out well for most of what the equation offers, but close to that infinity, an experimentalist ought to determine what is likely occurring. We shall see in later explorations of how STCs can distort in a way greater than what the speed of light gives us, suggesting a faster than light (FTL) quality.

So, Einstein used the same principle to outline what would happen to space itself as space-time is after all one thing. In order for space to change, we could form our own line of thinking. Especially from what we know of blackholes, we could imagine them getting bigger or smaller. If the result was bigger, they would be exponentially bigger making them far more visible than the other stars, not because of the blackhole itself, but from a larger event horizon and the matter getting sucked in. For a blackhole with the mass of thousands of stars and the added hypothetical quality of getting bigger from relativity, the blackhole would be exponentially larger. As this seems to be the opposite of what we find, logic suggests that the space itself is getting exponentially smaller.

I will discuss the relationship of time, space, matter, and energy later. For now, to keep the space shrinking instead of getting bigger, a similar equation for that of time is used, but modified just ever so slightly to decrease the value of space. This equation for space contraction is:

The only new variable here is l represents length. In this case, as v approaches c, we find that the value of l for length shrinks to a l' of zero. As we mentioned in holes, again experimentation will offer corrections to the actual value at the extreme of $v = c$.

So, as the spaceship approaches the speed of light, according from a reference frame of someone on Earth observing you on your fast-moving spaceship, they will see you as a shrinking or contracted in length and looking through a window, you would be moving very slow, almost (but not quite) as though you were in suspended animation.

Do note that I was a bit careful in not suggesting (yet anyway) that the spaceship itself, as in its mass was shrinking. It was the space itself that contracted and the spaceship mass is within this space. This distinction is quite important.

I have tried and explain that this is a real thing that is happening here but without proof, how do we know this is occurring in reality. After all, if it is just an idea, it is meaningless. For this, I am going to throw in another known principle that has also been established that the bending of time-space from this large velocity is additive in nature to the bending of time space caused by massive objects like the sun, neutron star, or a blackhole. How much this time-space bends is observed as the bending reveals itself as gravity.

We know that the gravity among the moon and planets are different based on the amount of mass they have. The same is true of the Sun as it is really massive.

The proof that Einstein needed to prove his theory came much later after he had published his work. He waited until a solar eclipse occurred. In it, he found that a far away star passing in back of the solar eclipse seemed to speed up its orbit as it got close to the sun. As it came out on the other side, it seemed to slow down back to its normal orbit. What was happening is the space just outside of the sun had warped the space to get smaller and the light from the star behind was following a straight path, but the path (space itself) was curving.

Science is great and all, but science itself is far less useful if we cannot find an application for it. So, I will list a couple of technologies where this bending of space-time gives us practical applications.

So, another test was devised to prove that time slows down as well, which the eclipse might not have directly proved. Two atomic clocks were used with one on the ground and another was placed on an airplane. From the airplane having a faster speed than the stationary atomic clock, the airplane landed and the difference in time passing worked out and demonstrated that the equation does in fact hold.

Having proof now, what applications can we produce. Here are two straight forward ones. GPS satellites are passing time-space differently than that of the Earth that is stationary. These differences in time-space allow for a triangulation of a ground-based GPS. It is now an application so prevalent that we now have it available on every cellphone today.

Another application that I have personally used is the atomic clock linked to navigation aboard a submarine. Having underwater maps of underwater mountains and such, the linked atomic clock picks up very small differences in our speed and direction and such that we can accurately navigate things like the mountains. Submarines do not have windows to see such things. Also, at least American submarines do not use an active ping to see the underwater mountain because that would give away its position. Submarine service is called the Silent Service for a reason. We even go to an extreme such that boots are not worn. We use sneakers because boots are louder than what we want. To be clear, passive sonar is also linked into the overall navigation system. With that, we can pick up the sound of shrimp on the ocean floor and that also assists us with the maps to keep us on track.

The really cool thing though is that traveling, even at low speed through deep waters, this still causes a time differential such that the atomic clock is only measuring with respect to the submarine. Intermittently the submarine has to go to periscope depth (the periscope has an antenna on it) and download the time and place it is with respect to the stationary observer on land. If we did not, after a two- or three-month deployment, our "known position" could be off far enough such that we would not have a real position of the underwater mountain and we could potentially crash.

In conclusion, not only is time and space dilation really cool, but it also offers real world applications. As much as we do know of relativity and its relation to gravity and such, it is currently one of four known forces; strong and weak force, electromagnetic force, and gravity, however, it is only the first three that we are able to link in with one another. Gravity seems to be on its own for now.

Gravity as the Odd Ball

As we look at how galaxies are interacting, gravity seems to be working counter than it should. Within a galaxy, we can tell that gravity is the same there as it is here locally. Instead of galaxies pulling themselves together as they should, there seems to be another force at work that we cannot see. For lack of a better word, we call them dark mass and dark energy. It seems to account for 90% of the mass and energy in the universe, but we have no way of detecting it other than watching the galaxies pushing each other apart. Even when we take into consideration that the universe is expanding and accelerating, that unknown force still leaves us at odds. There is even consideration that a fifth fundamental force might even exist. That is mind boggling considering it is so big that we are missing it. This is not the first time this has happened though. The Higgs boson, nicknamed the god particle was difficult to isolate in a particle accelerator. It is nicknamed the god particle as in it was god-damn hard to find. Once the Higgs boson was looked at in a wider view, it was finally found.

A similar oddity that I have heard has to do with our Milky Way galaxy. The galaxy is composed of a super massive blackhole at the center (many times larger than any regular blackhole) and several legs that kind of spiral around, being held in orbit of the super massive blackhole. The problem is that current models of known mechanics only allows for the galaxy to spin in a complete rotation before the legs should have separated and gone their own way. From my understanding, this also remains a mystery to science still.

Another fact that I originally found odd was the dimensions of the galaxy. In its circular orbit, the Milky Way is about one or two hundred lightyears across. The height is only one lightyear high. A professor explained it as the conservation of momentum that allows for those odd dimensions. I can buy that for now, though I have not personally modeled it.

Space-Time Curvatures

Space-Time Curvature (STC) is an old and established concept. Very simply, if we look at a star, it is massive and creates gravity. As we look at time and space, they turn out to be a single thing, a 4-dimensional vector. An advanced way of viewing this is a matrix form,

$$\begin{bmatrix} ct' \\ x' \\ y' \\ z' \end{bmatrix}$$

In simple terms, if a star has a high mass, it will take this single unit of space-time that is one thing, and it will cause a curvature on that space-time. As a result of this curvature, we happen to recognize it in physical terms as an increase of gravity. If we do the opposite, such as consider the mass of the moon compared to Earth, we find that this curvature is less, and therefore we see far less gravity on the moon compared to Earth. The Sun, Earth, and moon are examples of different STC.

STC are not solely due to massive bodies. If we took a probe in space and moved it towards the speed of light, we would also find that this high velocity also creates STC.

Of course, we could model STC in such a way that we consider the exponential scale from 0 to 90 degrees and apply some known examples as to where they would roughly fall. In the vacuum of space, the STC are straight and would fall at 0 degrees. The Earth and moon have little curvature due to relative low mass and be 1 degree. The Sun is pretty massive and might hit a 10. A neutron star is pretty dense so let's give it 50 degrees. The event horizon is very bent and would be an 89-degree mark of the 90 available.

With our space probe, we can get to 90% of the speed of light and also observe that this phenomenon also on its own can bend the STC at 80 degrees. If at 99% of the speed of light, the STC would be at the 89-degree mark, just like the blackhole we were considering.

What is more interesting, along with being well established already is the fact that the STC caused by massive bodies and the high velocity of the probe are additive in nature. This means that theoretically, we could take our space probe moving fast that produces a 50-degree curvature and fly it in the atmosphere of a heavy and massive star that produces a 50-degree curvature and the probe would have a resultant curvature of 100 degrees. This is a resultant of 100 degrees, farther than the 90 degrees that is indicative of the STC of the speed of light. This suggests that we can take our probe and even without achieving the speed of light, it is capable of showing us a result of the probe traveling faster than light FTL. While many might be surprised by this result, it shouldn't be because these principles have already been established for many years. I have yet to offer anything novel yet.

Next, I will describe how little it takes for STC to make a huge difference. We even have examples of this that you can do easily at home. Earth has a weak STC as we have already established. We can consider the STC on Earth at the elevation of our feet. This STC offers a curvature such that the gravity acceleration is $9.8 \frac{m}{s^2}$. We can also go high in the sky and before we hit an elevation where it might orbit the Earth, we can find an elevation such that the gravity acceleration is $2 \frac{m}{s^2}$. There is obviously a difference in these two STC. If we dropped a marble from the higher point, it would begin to fall, but with a slower acceleration, but as it got closer and closer to Earth, its acceleration and velocity would increase until it hit the ground where it was picking up velocity and acceleration on the way down. For simplicity, let us ignore the effects of air friction as well as terminal velocity. We could easily use a simple device as an accelerometer at the lower and higher elevation and determine the degrees of difference in STC. We could also use the same really accurate accelerometer to look at the difference in STC between the distance of your feet and your head. We would find the degrees of difference between your feet and head is very small, but with an accurate enough device, there is a small difference. It is enough such that if you take a marble at the level of your head and drop it, that minute difference is enough to build up a pretty good speed. We can further explore the idea of taking the marble to the top of a 100-foot building and dropping it there. There is a slightly larger difference in STC, though on our 90 scale, we might have to look at quite a few digits beyond the decimal point to actually see the difference.

The big take away here is just by creating a really small differential STC, even this small acceleration of $9.8 \frac{m}{s^2}$ is capable of creating a high velocity just seconds into a fall. Ignoring air friction, imagine the speed that could be built after a few minutes (think sky diving).

Now, as we go back to the idea of a space probe, imagine if at the front of the probe we could create that really small curvature through some means in the front of the probe and we could straighten out the STC in the rear. That small curvature would be "free falling" in the forward direction and continue accelerating. At this acceleration, it would not be relatively a long period of time before the probe built enough velocity before we would get to the 50-degree mark on our 90-degree scale, where we could begin to see some relativistic phenomena. Further, we could navigate it into the atmosphere of a star for an extra boost in STC. While fascinating, so far, all of this is known science, just not really well explored.

4-D Curvature

Touching back on special relativity, we have:

and

For the variable of v, this velocity is a vector, meaning that it not only has a speed, but also a direction. I am going to break with mainstream ideas here. Currently, mainstream thoughts for a spaceship traveling close to the speed of light would say that since v is a vector, the length of the spaceship is only contracting in the direction of travel. I propose that instead, all three units of length are contracted. Do remember, we are not shrinking the mass, but instead, we are shrinking the space that the mass lies in. Further, the structure of electrons flowing in orbits or shells around the atom, this is a phenomenon that is occurring in 4-D space-time. While this might be a weak argument, we have another stronghold. We have real observations of the curvature around the Sun. This is bending space-time in all four dimensions, not just time and one dimension of space. And finally, our knowledge for the equation of gravity:

$$F_g = \frac{Gm_1 m_2}{r^2}$$

It is fundamental that gravity must work in 4-D time space. If space begins to curve in one dimension of space, the remaining two spatial dimensions must follow suite.

Don't get me wrong, there would be an added bonus if it were curved in only one spatial dimension. This would allow us at some point in time to create a ray gun of sorts, pointing a gravitational burst in one direction, making for a fancy gun. However, among any phenomenon observed, gravity, which is the 4-D curvature of time space, it always curves time to shrink and spatial coordinates to contract in all directions.

Structure of atoms are circular

Sun gravity

Gravity can only travel in 3D space

The Sun's Magnetic Poles

Going back to the basics, I think that we can look at an old experiment from Einstein for some new equations. It has been suggested that magnetic fields straighten out STC. If true, this would be a practical and low-tech method of assisting in a warp bubble such that we create a curvature in front of our probe and use the new tool of magnetism to straighten out STC in the rear, offering an even larger differential STC, resulting in relatively easy modification of STC and higher accelerations.

I was watching a show on astronomy and very intrigued by a technology I saw. An astronomer put a metal plate over a telescope lens, but had drilled holes in it such that it would block out the Sun, but allow a star near where the Sun's atmosphere would be. We could therefore repeat the original experiment Einstein used without having to wait for a solar eclipse.

This method could also be used to look at stars just beyond the atmosphere of the Sun above the magnetic north pole and magnetic south pole during the event of the Sun changing magnetic poles. For some reason, which I am not sure if anyone understands, roughly every 11 years, the magnetic poles switch their polarity. What I anticipate is that during this transition, as we monitor the stars behind (and slightly above) the Sun, we shall see a strong distortion in the path of those stars beyond. More importantly, data gathered would give us a better idea of the mechanism that is occurring during the transition. If the STC are actually straightened out, we shall know by how much. More interesting is that as the phenomena takes place, we could record it as a function of time, thereby gaining insight into not only how much the shift occurs, but also in terms of the changing magnetic field, which might have its own implications. This would show us how to use magnetic fields in the rear of the probe to give far more bang for the buck in the differential STC for a probe using low-tech. A challenge that will persist though, the magnetic field created by the Sun is probably far greater than any type of magnetic field that we could create on Earth.

The Bumper Effect

This is a far more difficult concept even for physicists to get and I am not doing much justice to the idea because I am explaining it qualitatively rather than quantitively with equations. Here goes anyway. Using LT equations, we can plug in a speed and get two results, a non-local length and time dilation. Getting to that relativistic speed required an acceleration though. If we consider a pretty long ship let's say 1000m long and break it up into 3 sections at 333m each, forward, rear, and in the middle, already going fast and getting faster through acceleration, the three sections are not moving at the same speed. I call this the "bumper effect." You can hold a block in front of you and move it in front of you until you convince yourself of it. As this ship is moving in front of you (and accelerating), the whole ship is continuing to shrink. I am going to give an arbitrary point it shrinks to (while I believe to be correct, I am still calling it arbitrary) towards the center mass of the ship (which we will say for now is in the exact middle of the ship). The middle section of the ship has the back half shrinking towards the middle and the front half shrinking towards the middle. In the front section of the ship, that 1/3rd, this whole section is shrinking backwards towards the middle of the ship and the whole 1/3rd of the rear section of the ship is shrinking towards the middle of the ship. So, the exact middle of the ship is moving at velocity v, but increasing because of the acceleration, nothing funny here. The whole ship shrinks non-locally according to LT equations. So, the front third of the ship, what speed is it moving? It is moving at $v - at$, where v is the speed at the center minus the a (acceleration) times t (time), making it not go as fast as the center. The rear though is moving at v, but also moving even faster because of the LT shrinking towards the center. This makes the rear bumper going faster than the center and even faster than the front bumper during acceleration.

To add bells and whistles to the probe and how we might get more insights, let us consider making the rear significantly less massive than the front. It is going to just be a probe after all. The middle, also not very massive and we just want it to be rigid to hold the ship together. The front holds the fuel source and is very massive. The goal is to keep the center mass in the front of the ship as long as possible. As the ship accelerates, the acceleration will cause the different parts of the ship to contract at different rates. Non-locally, this will cause the center mass of the ship to appear to shift towards the rear (relativistic mass) as the rear is moving at a higher velocity than the forward. The increased speed in the forward will cause the energy density of the fuel to increase (being one solution to the infinite energy problem). As the rear accelerates and has a higher speed than the middle and even more than the front, its shrinkage will be greater and relativistic mass increase, causing the center mass to shift towards the rear.

Now these high speeds and high velocities are bending the STC which is all geared towards creating the STC greatest on the probe near the rear. Now let us add some more bells and whistles. More rear than the probe near the rear, we are going to add a magnetic field. As adding a magnetic field will straighten out STC in the rear. The reason for this is to have a differential from the forward and the rear. This warp bubble being more curved in the forward and more straight in the rear creates a large differential in STC. Unlike the really small differential on Earth, this will provide a very large differential in STC aboard the ship. This causes a "falling" effect where the ship further falls (or accelerates) faster towards the forward of the ship providing propulsion that appears as a mass falling in gravity, so I call this gravitational propulsion.

As the forward energy density increases, this will allow for an "infinite energy density" that should overcome the problem of needing an infinite energy density.

More so, we can take the ship (and probe) moving really fast and eventually travel in the atmosphere of a star, further bending the STC beyond the 90-degree mark, giving the effects (maybe a worm hole) of traveling faster than the speed of light without having to actually travel the speed of light. True effects are again for the experimental physicist to observe.

Here is why I believe a worm hole would occur, along with more of that radiation. We remember that the back of the probe is moving faster than the forward. While the STC of the forward is not at the 90-degree mark, the STC for the rear is beyond that of the forward so it really does not make sense that the rear has moved past the forward bumper so I can only imagine it entering into some sort of hyper space. The center is still rigid holding the probe together with the front to make sure it does not fall apart. The forward moving sub-light speeds, this forward section will now act as an anchor for the overall ship to remain in normal space-time non-locally. So, we have pushed a probe past light speed, at least effectively, being faster than light speed FTL.

On the rear end with the probe, lets add some more bells and whistles. On a forward section, if we can imagine a rotating section of a cylinder such that it is rotating sideways to the forward, a very fast-moving rotation can also further curve STC. The part more forward will assist in the STC while a magnetic field in the rear of the cylinder will straighten out the STC. This will further create a local bubble within the probe with a larger differential in the STC. In the forward end, we can also imagine a metal ball being induced with magnetism and spinning in a circle that would strategically be rotating in the y-z direction and adding further magnets such that the main magnetism is on the rear end and the additional magnets are canceling the magnetism in the forward. Such setup would further create the differential in STC. All of this engineering is giving us more bang for the buck.

Solar Collectors and Solar Cells

I will use this design in the book later, but I wanted to take a preemptive strike at it. There is a difference between a solar collector and a solar cell. The solar cell works by using a particular material exposed to the Sun and a particular range of light frequencies has the ability to give an electron the ability to move from its spot.

Here is a curiosity I have, and have been unable to reconcile myself. If we put a thin transparent film over the solar cell where it would cause a higher or lower frequency of light to change to a higher or lower frequency the solar cell would work, would this in fact allow a larger range of spectrum of light to make the solar cell produce more energy? The simple answer for me is I do not know, but remain curious.

Next, a nuclear reactor and a microwave have much in common. The microwave cooks food by sending microwave energy through the food and absorbed by water. This extra energy causes the water molecules to be excited and bump into each other and this causes a friction which is in turn heats food. The nuclear reactor is similar, but uses the radiation from uranium instead.

A nuclear reactor has a loop of water. Within this loop is also a turbine. I am really oversimplifying this, but this heat causes the water in the loop to move, which in turn causes the turbine to spin. Just as the windmill turning generates electricity in a generator, this nuclear reactor is doing the same.

Looking at a solar collector, we are going to keep the loop of water and the turbine, but instead of heat being caused by a nuclear power supply, we are going to use many mirrors that take light from the Sun hitting it and focus that light up on a tower. These mirrors are automated to move in such a way that each of them adjusts for the Sun's movement and ensuring that light is still hitting the tower. These mirrors are pretty inexpensive, being flat mirrors.

From the tower, there is a pretty large mirror that is parabolic in shape and looks like a satellite dish. This is not a cheap mirror because of the precision that it must be curved just right. Its function is to take in all of the light from all of the other mirrors and focus it into a single tight beam down onto a single point onto our loop of water. This process is very much like taking a magnifying lens and focusing the light from the Sun to burn an ant. FYI, don't do that. It is cruel. The outcome of this focused light on a point of our loop heats up the water and performs the same function as the nuclear reactor part onto the loop.

Early on with the building of these solar collectors, the basic problem came up that these reactors were only working during the day time. Trying to collect enough energy and storing the energy in batteries was not practical nor economical. The work around was pretty ingenious. Rather than heating water, the design was altered to heat up salt into a molten state. As this was a much higher temperature and the specific heat that it could store was greater, the loop acted the same during the day time but had enough energy throughout the night to keep spinning the turbine. This allowed for power to be delivered 24/7. Those are the basics. I will cover more advanced bells and whistles later.

The Evolution of Batteries

It wasn't that long ago that a decent size battery was no different than the battery in your car (primarily being lead-acid). As solar collectors, solar cells, windmills, and even power generation from the ocean waves have evolved and creating new types of power generation, the ways of storing this energy is also becoming its own field.

One of the types of energy storage that I have found most intriguing lately is taking an underground cave system, ensuring it is sealed, and from there using any kind of power generation to pressurize the air within the cave just as a simple air compressor. An air compressor takes pressurized air and powers pneumatic tools. If you have a larger tank, you have more time to use that tool. Similarly, if you have a really large sealed cave, that would act as a massive air compressor tank.

So, any kind of power generation can act on the compressor side, whether coal, wind turbines, or solar cells. Solar cells are down throughout the night so this storage method of energy is perfect. What is even better about this method of storage is that we do not always have to go looking for these hollow caverns. People have been creating them artificially for a long time now. Mining operations that have come and gone have created many of these artificial caverns for a long time now. A closed mineshaft also usually leaves very cheap real estate as well. It appears that this is underutilized.

Datacenters seem to choose their locations for specific reasons. How much risk does the location have for tornadoes, Earthquakes, loss of utility power or water, latency, etc.... Datacenters use a massive amount of battery power that is only intended to power racks in case of primary power loss but only for a short duration. It appears that this pressurized cavern would offer a very cheap and reliable method of "battery backup" or even an extra leg of latency. For some end users and applications, latency and speed are less important, such as storing medical records, but still require access to documents at all times. Having to wait 5 seconds for retrieval of data is not problematic. Further, data is often stored in multiple locations just in case one rack goes down. This would provide a vault to store data in the worst-case scenario that all of those servers went down that is again using a cheap energy backup system.

Electric cars seem to have taken off and left hydrogen power cars out of the game. The federal government has also provided a boost in charging cars. While it is high minded to want to implement cars being electric, there are issues that seem to not be adequately addressed. On the street I live, the power lines not only to the street, but also from the street to my home are going to be inadequate to power electric cars. I would be responsible for the additional costs of more electric lines running from the street to my home breaker box (a new one that can handle more power). Currently, if I run a hair dryer and an electric heater, I am already tripping a breaker. How is the power company looking forward to these additions even on their end? It is currently not happening where I live.

Thinking that these electric cars are better for the environment and addressing climate change is a misnomer. The power going to these cars is still by burning coal primarily, then converted to electricity, which involves a loss in efficiency, using more energy to power the car.

As a middle ground to solve the problem, hydrogen, or even the hybrid of a hydrogen and electric vehicle is probably a better solution. Hydrogen can be stored in tanks and moved anywhere, including by train, whereas attempting to move electric current further on a transmission line to a more remote spot (including gas stations that can charge electric cars) involves transmission losses. Hydrogen does not loose energy via transport or moving to remote locations. To that end, for these remote places that we are using EV's, it might be best to use a renewable source of energy to store hydrogen, transport to a remote EV charging station and power the charging station with a hydrogen engine. Of special note, as mentioned previously with the sealed caverns storing energy, these are often in remote locations so integrating these into our network of power for EV charging would also be advantageous.

Endothermic Chip

Energy comes in many forms, whether light, electricity, heat, or sound. Technology surrounding them is a matter of using processes and the inherent properties of matter to exploit these. For example, only certain materials can utilize the photoelectric effect to capture light for solar cells to produce electricity. Just as we add slight modifications to prescription eye glasses for different effects, like having them turn into Sun glasses in the presence of certain light, we may even be able to add such layers to solar cells to increase the range of frequencies to better maximize power production and minimize surface area, increasing the efficiency of solar cells. In a way, we already have an old technology that does this, ignoring the idea of solar cells already, think solar oven. As is similar in a solar oven, a solar collector uses mirrors over a vast surface area and reflects it to one point in the center, which is then focused with a parabolic mirror onto a loop of liquid which in turn turns a turbine and generator to produce electricity. Modifying the technology just a little bit uses salt that becomes molten so the heat it absorbs will keep it liquid throughout the night, which allows for the generation of electricity using the Sun, even during the dark hours.

The process of developing most any new technology works in a similar process. We begin with a simple principle where nature cooperates with us and we fine tune it, using more principles such that the technology will work more efficiently. As another example, let us consider that we want to create a power generating technology that does not use any land. We are creating a self-imposed restriction. Very simply, we decide that we are going to build a platform on the ocean (avoiding land use) and place a wind turbine on it. Since we are already doing the work of placing a platform on the ocean, why not just put some solar cells on top, collecting electricity from the Sun as well. It is just a little more money invested into what we are already spending the bulk of the money into the platform and wind turbine. The additional solar cell is not going to interfere with the operation of the wind tunnel. The point here is to look at exploiting natural phenomenon towards what we want to get out of it.

Now, let us consider a phenomenon that we have been exploiting for a long time with another goal in mind. I work in a datacenter and as is inherent to a datacenter, our computers generate a massive amount of heat in which we even separate computers in a cold and hot aisle for the purposes of keeping our computers from overheating. If we recycled that heat during the winter to heat business offices, we would be doing ourselves a great service. I have a different proposition for that heat though, towards an electronic device that does not yet exist.

Let us consider the old technology of the electronic components, the diode and transistor. Long ago, we used to use vacuum tubes in order to bias electricity to only travel in one direction. Then, with particular chemicals, we found that we could bias electron flow in one direction with a far lower voltage with a couple chemicals in a diode. For high voltages, we still use vacuum tubes, but for some circuitry with low voltages, we now use diodes using two types of materials. It wasn't long before we recognized that we could use three distinct layers (along with a concept called doping) that allowed us to switch the electricity to flow or not flow quickly. It wasn't too long after that we realized that we could now use this in terms of logic gates and microprocessors. A microprocessor is a collection of many many transistors. As time has progressed, we have developed processes of how to get more and more transistors onto a single microchip and get them closer together to increase their speed.

There is a natural by-product that occurs in the development of more dense microchips that works against this newer technology, heat. They now generate so much heat that a heat sync has to be attached to them to pull the heat away from the microprocessor and have a fan blow over that so that the microprocessor does not burn itself out. What if we could modify the technology instead to remove that heat in a different way? In a datacenter, we use a massive amount of energy for the sole purpose of removing that heat, whether that is in air handler units, or chillers, or cooling towers.

Using the same method of developing new technology, working on natural phenomenon, I propose going back to the drawing board at a level of the individual transistor, but rather than making a switch, we use chemical processes to absorb heat, and in some way, make that wasted heat energy do something else. This is the beginning of the endothermic chip.

Most chemical reactions are exothermic in nature. One simple example we might say is the combustion of gasoline in an engine. Gasoline is combusted and we get a new set of chemicals and heated gasses that have a higher pressure and temperature when contained, moving a piston in the engine. Other chemical reactions of two or more chemicals often work together forming new chemicals along with heat. There are several (and not near as many) that instead of creating heat, they absorb ambient heat, and require it, for the formation of a new chemical. When this formation takes place, the new chemical feels colder than the previous chemicals before formation.

This gives us a new phenomenon in nature to attempt to exploit. I must say at this point that the technology has never been attempted, at least in my researching looking for such technology. Exploiting what we know from a transistor, in this new case, we are looking for two chemicals that instead of acting like a switch, in nature they come together and create an endothermic reaction, absorbing ambient heat. The problem that we initially run across is that if we created such a device where these chemicals come in contact with each other, they would simply form a new chemical and we only got to use the phenomenon one single time. We are wanting this to be a continuing process that occurs over and over without the formation of a new chemical.

This being the case, we are going to have to put something in the middle of these two chemicals firstly for the purpose of completing the formation of a new chemical. Secondly, this barrier has to conducive to the valence shell of electrons between the two to appear as they are interacting similar enough to the formation, but without actually touching. In a transistor, special chemicals along with doping allows for this to occur, while still only allowing electrons to flow in one direction.

The good news in researching this new device is that unlike reactions that are exothermic, there are very few combinations that are known to occur as endothermic. This significantly reduces the number of trials possible. On a side note, if this technology seems to work even just a little bit, far more research will go into finding other endothermic reactions that we do not know of today.

Going back to our endothermic chip, we know that we have to use sets of chemicals that produce the function of endothermic and now we just need to find our barrier. The bulk of the theory resides in the fact that how valence electrons from the chemicals form is more important in creating the by-product more than anything else. It might be that this barrier is going to be a different chemical altogether than is currently used in transistors. For getting materials very close together, if this is the needed requirement, we might use gold foil. Gold foil can be flattened to such a degree that the foil has a thickness of a single atom thick. It has an additional property of allowing electrons to flow quite easily, as it is a great conductor. Now, assuming we have a rough sketch of what the technology looks like, I will now explain what the operation would look like along with what we might find instead.

Two chemicals that are naturally endothermic in nature are separated by a thin barrier. It would likely need some sort of jump start, possibly by getting a jump start of electricity on the barrier. This electromagnetic field would allow for the valence shells of the two endothermic chemicals to interact in a way that attempts to complete a formation of the new chemical, but far enough away that the actual formation cannot actually occur. In this pseudo-reaction, this would be enough to cause the occurrence of drawing in heat in an attempt to complete the reaction. Very simply from the principle of the Law of Conservation of Mass and Energy, as this energy in the form of heat is drawn in, it must go somewhere. My main consideration is that it will also convert this form of energy into electricity (though as I will explain, it might be in the form of light). As there will be a bias in the direction of flow of these electrons, it is that bias of direction of electron flow towards a transistor gate, which is now powering the transistor. As that heat energy is absorbed, it must expel the energy in some form. It is unlikely, though possible, that it would convert that heat energy back into heat energy. Certainly, as we vary the combination of endothermic materials used, we would find that we will get the converted heat energy into not just electricity, but maybe other forms of energy, like the formation of light. An LED (light emitting diode, with diode being the key word here) is the utilizing that natural phenomenon of converting electricity to light. The LED is simply an extension of the technology derived from the simple diode that originally only allowed for a small voltage to have current flow in one direction. Even if this endothermic chip amounts to only creating the property of light, it would have the distinct advantage of requiring almost no initial electricity and continue in its process of creating light while drawing in ambient heat as the main power source. It would be more beneficial to keep on such a light, being more efficient than actually turning it off and then back on.

As we find the right set of endothermic materials and a proper barrier in the middle, the design of this type of transistor can be reduced into the size of the transistor used in a microchip. For such combination that naturally draws in ambient heat and converts it into electricity, many of these can be strategically incorporated (miniaturized of course) in order to collect the heat generated by the normal transistor on the microchip and convert that to the electricity to power the normal transistor. The resultant effect would be a microprocessor that is not only self-cooling, but also self-powering. For a datacenter, this would not only remove the need for cooling infrastructure and electricity to power the racks, but it would reduce the surface area of the building to a quarter of the required space. Cooling structures take up far more space than the rack zones itself. Further, the electrical lineup and switchboards require a lot of space and equipment to provide the clean power needed for the power sensitive racks. For battery storage alone that is used as a backup, the reduction in necessary batteries can greatly be reduced and still provide a longer backup time if necessary. For what would normally require 18 diesel generators for a building would reduce to a single generator with much to spare.

So the holy grail is electricity generation using this component, we can also see the usefulness in such combinations that also create light, vibrations, or even sound for a new type of speaker maybe.

Propellent-less Rocket

For this, I am going to begin with a very classical design and then carry it into a more formalized format of utilizing plasma in a similar way. Let us begin with a probe in the vacuum of space that is initially stationary.

From all known accounts, we require a fast-moving gas to be expelled from the rocket for propulsion. This boils down to the conservation of momentum where the equation making it work is:

$$m_1 v_1 = m_2 v_2$$

On our probe, it is pretty bare at the moment. The only special component it has is a loaded gun attached to the body of the probe. We can even give it a length inside, say 10 meters in length. If we pull the trigger, the initial motion of the probe moves the probe forward. When the bullet reaches the rear, we are working with an equal and opposite reaction such that when the bullet strikes the rear, that mv factor of momentum gives an

exact value that causes the probe to move back to its exact starting place. This conservation of momentum is not something that can be circumvented.

We can however use the same equations, with a little tweaking to get a bit of a different result. To begin with, lets assume that the body of the probe is bullet proof and not going to get a hole in it when shot at. Next, at the rear, there is a metal plate at an angle such that when the bullet strikes a plate, the bullet is deflected at an angle of 90 degrees and hits the haul of the probe. Now, some of the energy, when hitting the plate, will cause some of that momentum, but as deflected and hitting the wall at 90 degrees, the probe will move in a resultant direction of to the rear and also to the side.

Now, let us use two guns and two plates. The plates will deflect one bullet to the left wall and the other to the right. If we assume the two guns now create a hypothetical force of 20 Newtons of force in the forward motion for the probe and then when two bullets hit the plates, there is a combined force of 10 Newtons towards the rear. The bullets then travel and hit the right side of 5 Newtons and the left side of the hull at also 5 Newtons. For the bullets that hit the left and right side of the hull at the same force, but equal force and opposite direction, these forces equalize each other. The resultant of these vector forces is 20N-10N=10N of Force in the forward direction. It is not really that fancy of physics going on here. If this were a black box though, let us say disguised as a comet and this process happens repeatedly, it would seem that something really funny is going on because this "comet" seems to be able to accelerate on its own. This setup could even be further modified such that the plates shift their angle a bit if desired and now not only do we have a "comet" that can accelerate, but can also seem to adjust its direction with no known method as there is no propellent.

There still remains the task of using some mechanism to get the bullet back to the forward with minimal force in order to recycle it and do this repetitively. At this point, however, we can start to use an internal propellent, say a plasma, along with our plates for deflection. The resultant would be similar, however, working with gasses, this gives us an added advantage of recycling our gas. By simply creating multiple chambers that takes our high-pressure gas and offering several chambers with weaker and weaker pressures, the high-pressure gas will want to move to the lower pressure chamber on its own, moving to ever decreasing pressure chambers until it has returned to its original spot. This should overcome the problem of recycling the internal propellent.

Overcoming the challenge of momentum through deflection does allow us to vastly reduce the requirement of huge rockets attached to our small space probe or shuttle (including getting off of Earth), but we are still leaving out a major component. If the original probe wanted to hide all of its secrets to just look mysterious, it would have to have stored a lot of energy internally to continue to look like a black box. Instead, let us say that we want our probe to overcome the challenges of lifting off the Earth and traveling through space without the propellent, we need to have a large supply of raw energy. This is where we are going to fall back onto basic known technology to fix our new problem of requiring a lot of energy.

I previously mentioned a difference between a solar collector and a solar cell. A solar collector is not much different than using a magnifying lens, condensing the light with a surface area and directing it to a simple point to cause a fire. Modern concepts of this has been to put such a mirror in orbit of Earth, convert it to microwaves and send it to a particular spot on Earth to power a loop of liquid, thus turning a turbine for energy. Such basic technology easily be improved by using a network of relay mirror satellites such that as one collects light and sends some of it down to the desired spot, it also sends a good portion to a network around the globe where another satellite is in darkness, allowing some of that light to be sent for a spot without the Sun. This network would overcome the challenge of using light day and night.

I propose a different arrangement which could easily solve the world's problem of energy and climate change overnight when the system is turned on. We will keep the network of satellites, but instead of collecting light, they remain solely responsible for dividing light and sending it to the proper location on Earth.

As the Sun expels light, there is one primary loss in energy. As light comes from the Sun, it spreads out at a proportion as it gets further from the Sun. To the opposite, as we put a mirror closer to the Sun, we bypass the spreading loss and get an exponentially larger amount of energy available. To this end, if we had one large mirror that were responsible for the collection of light in a closer orbit to the Sun that could condense it to a tight beam and from there send it to a network of satellites around Earth, a single large mirror could easily provide all energy needs of the planet. Again, the network of mirrors around the planet would not be for collecting light, but just a single tight beam as well as sharing the portions of light to collection points around the globe.

We run into the obvious problem of the satellite mirror is now orbiting the Sun, so what about when it is not in view of Earth? Well of course a few network satellites around the Sun as well solves the obvious problem.

How does this solve the inherent problem to our original space probe? When we consider the outdated space shuttles, it is only a tiny portion of mass compared to the overall mass of the shuttle and rockets of fuel used to get it into orbit. The amount of raw energy needed to get just a shuttle off the ground rather than combination of a shuttle with its rockets and fuel is quite significant. I am not sure if it is correct, but the statistic that I heard is that during the launch of a space shuttle, during that time of getting off the Earth took as much energy as all of the United States itself was also using during that launch. If that energy needed was only the much small shuttle and not the rockets as well, it would require far less. To that end, if we considered a similar network of light beams highly condensed from a closer orbit of the Sun onto the shuttle itself, this would be a gamechanger for space flight. It would be more significant than reusable rockets as the entire rocket itself would become obsolete.

Further, even when a probe is traveling and accelerating (even not needing a gravitational boost from a planet) through space, as the tight beam of light from the Sun keeps powering our probe, we would not be counting on a velocity to take us out of the solar system, but instead, we could continue accelerating so long as the beam remains tight and focused on a point that the probe can accept the beam.

For further consideration of power on Earth, there are some fundamental challenges that would still come into play. They are not beyond current technology though. The first problem is that focusing this beam of light onto Earth is rather dangerous. If there were any sort of alignment issues with the beam, this has the potential of cutting across the US and leaving a powerful laser burning a large swath of land. There would have to be a mechanism in place such that if something alters the beam at any of these satellites, the beam would have the ability to shutdown or redirect in a safe direction. Second, it is likely that it would be beneficial to have the collection node on the surface of the ocean, far away from people.

This introduces a problem of transmission losses. The further that electricity has to travel, the more losses that it will incur. So, how could we take an unlimited power supply and move it to the middle of the United States? On the ocean, we have an invaluable resource, water. With enough energy (which we have much of now), using a good amount of that, we can easily take apart large quantities of water apart and isolate hydrogen. Hydrogen can be condensed into a liquid form and travel anywhere without losses in tanks. I have considered the possibility of creating pipelines across vast distances and this would in fact be more efficient. As we see with modern oil pipelines, we have a poor record of pipelines bursting. If we were wise, for oil pipelines, there would be a secondary pipe that contains the inner that actually holds the oil, such that if there were a burst to occur, it would all be contained. Current users of oil pipelines probably view the cost of environmental disasters and cleanup cheaper than encasing the primary pipeline and that is unfortunate that saving some money is cheaper than an actual solution. To that end, the secondary piping for a hydrogen pipeline is not only there to contain a leak that can be detected, but hydrogen is also highly explosive. It is therefore best transported in reinforced tanks rather than a pipeline.

E=mc²

The actual equation for $E=mc^2$ is $E^2=(mc^2)^2+(pc)^2$. The only new value here is p that stands for momentum. For a nuclear reactor, the very small mass that is converted into energy. I heard of a city on the east coast that got hit hard by a hurricane. The Navy hooked up two submarines with their relatively small nuclear reactors to city utility power that had gone down. Those two reactors powered a large city.

What is most interesting with the equation is the relationship between energy and mass. When I was growing up, I was taught in school of the Law of Conservation of Mass. Separately, I was taught of the Law of Conservation of Energy. $E=mc^2$ throws both of those laws out the window in exchange for the Law of Conversation of Mass and Energy. The conversion from mass to energy or vice versa no longer violates the new law.

If we consider $E^2=(mc^2)^2+(pc)^2$ such that there is no movement, $p \to 0$ and thus the term $(pc)^2 \to 0$, resolving our equation back to $E=mc^2$. Similarly, if we instead consider where $m \to 0$ and thus the term $(mc^2)^2 \to 0$, leaving the equation for a photon as $E=pc$, the energy of a photon.

The difficult question to answer now is does a photon have mass like the equation seems to suggest and why? For the equation $E=mc^2$, light is simply energy so it appears that maybe if we concentrate a lot of light in one place, we might be able to detect gravity from it. If we wanted just a small amount of mass that is simply detectable, we might be able to just rearrange our equation for what we are aiming for as $m=\dfrac{E}{c^2}$ and we have something to work with. However, this is not the case. Here are a couple of reasons that help us determine that.

If light had even a tiny amount of mass, that would imply that if it traveled over a really long distance, there would be the slightest of a spatial curvature. All mass creates this curvature. If light had the slightest of mass, and if we looked through a telescope for 1 billion lightyears away, we would find that that slight curvature in space would scatter all photons in different directions and we would not ever be able to see such a star with the telescope. Another consideration that I am not sure has been tested or not, but is rational is that the Sun is constantly converting some mass into energy in its fusion process. 4.26 million metric tons of matter is released in the form of energy per second [3]. Getting a "proper reading" of the Sun with its change in momentum and gravity over larger increments of time, this would tell us whether the light has mass or not. There would be a relatively consistent amount of light in all directions that are within the solar system. This flux of light could be summed up and show its own value of gravity in the solar system as different values depending on how far you are from the sun. By "proper reading," this is a slightly tricky observation to make. Taking into account the gravity that might come from the flux of light is not terribly difficult, however, the helium that is produced in the Sun from fusion, being denser naturally, that would influence the gravity that is trying to be observed. To that end, it is not one simple measurement, but must also take into account the mechanics of the fusion process itself.

This leads us to another consideration. If we postulate that the light stars give off is a continual process of losing mass over time, the Milky Way has a hundred billion stars and is losing the amount of mass from one star multiplied by that one hundred billion. Later, we will cover some of the complicated orbiting that occurs because gravity only travels at the speed of light.

For now, we can consider that the Milky Way is losing vast amounts of mass and the energy is increasing as a result. To that end, consider an ice skater. If he or she is spinning and has her arms out, she will have a set spin. However, having that same energy, she brings her arms in, she spins even faster. Applying this to the Milky Way, energy increases and mass decreases (keep in mind that the rate of loss of mass can be considered fairly consistent in such large terms), the velocity of spin of the galaxy ought to consistently increase. For matter that is on the outermost part of the galaxy, it ought to continually loose mass. This is because the small amount of gravity from the supermassive blackhole in the center is being overcome by the outer mass having a greater velocity that should allow it to escape. More on this mystery of how the galaxy should or should not be stable shall come later.

Birth of a blackhole; inside a blackhole and neutron star

Gravity difference in solar system

Milky way looses mass over time

$E = Ymc^2$

$$p = Ym_{relativistic}v$$

Relativistic mass

Relativistic Mass

For relativistic mass, I am not sure if my personal definition matches up to that of mainstream definitions, but I will present my own. Here are two equations:

$p = Y m_{relativistic} v$ and

For a simple case, let us say that we have a fast-moving spaceship that is moving at a high velocity (near the value of c)

and momentum and velocity are constant. This also gives us a constant value of gamma. Simple manipulation of these two equations reveal that the relativistic mass that is left must be higher than the mass if it were stationary.

This is odd though. We know that the Law of Conservation of Mass and Energy prevent mass from just coming about from nowhere. And, it is not. As we look back on special relativity that we have considered before:

We take note that the mass that is constant is within a unit of space that is shortened with respect to a stationary observer. Then we use our equation for gravity:

$$F_g = \frac{G m_1 m_2}{r^2}$$

For the value of r being a radius, we substitute in the value of l'.

From this, we get a consistent amount of mass with a shortened radius, and this produces a larger force of gravity F_g according to a stationary observer.

So, of particular note, it is not the mass of a fast-moving spaceship that is increasing, but the gravity according to a stationary observer that increases.

I am not sure I have adequately expressed the relationship between mass, energy, time, and space yet, so I will expound on that here. Time and space creates a curvature that tells mass and energy how to move and at what pace. On the flip side, mass and energy tell time and space how to bend. It is an intracity each set has its own power over the other set.

Quantum Entanglement

Quantum Entanglement is a difficult concept with many applications down the road. It is also extraordinarily difficult to explain. Einstein who helped develop it struggled with it a lot, calling it "spooky action at a distance" and tried in many ways to disprove it. While almost every time people would side with Einstein, this was one of the rare instances fellow scientists would disagree with him. Rather than attempt to explain this with difficult equations, I am going to hit on most of the concept qualitatively. I am also going to loosely use the word of *measurement* and observation and I am going to completely ignore their definitions because I struggle to define them for myself. As far as the means particles are *entangled*, I will give a vague definition as well and not going to well define this either.

Let us consider two particles that each have a spin, either plus or minus. Next, by means of maybe getting these two particles together really close for a short amount of time, this process will cause them to go into a quality called quantum entanglement. In this new state, we find that both particles now are in a state that neither has a positive nor negative spin. It is a third state that is neither positive or negative. We can even keep one of the particles local and send the other very far away, for example one lightyear away. After both particles are in place, we can perform an observation of the local particle. The observation causes the local particle to immediately take one of the two spins. Let us say we do our observation or measurement, and we see that the local particle takes a spin of minus. Regardless of the distance between the two, that second particle (which we said was a lightyear away) will immediately take the opposite spin. What bothered Einstein here was that somehow information is traveling faster than light.

This has some serious implications. Here is very well one of the strongest that we can consider. Let us assume that we have a star that is one billion lightyears away from us. From that star, a light wave or photon from the star and travels to a satellite in orbit of the north pole of Earth. Also, we have another photon that leaves the star and travels on a path with the original photon for a while, giving them proximity for long enough that they become entangled, but then diverge to very different places. When one photon hits our satellite after a billion years of travel, the act of observing the one photon "collapses the wave function" and causes the first photon to show that it has a minus spin. At that very moment, the other photon that can be any distance apart now must immediately assume the opposite spin, being plus. During the whole time of traveling for a billion years, it did not have this quality of spin. It was only when the observation was done that both assumed a spin.

Having acted faster than the speed of light, working instantly appears to violate the universal speed limit of the speed of light. There have been attempts of a work around such that since the particles themselves traveled over the time period, maybe they had their spin during the journey. This appears to not be the case in reality. Reality suggests that what is really going on is the collapse of the wave function is in fact occurring instantly. We could modify the above scenario slightly and say that particles become entangled just shortly after leaving the star and then travel independent paths for the billion years. Our act of observation a billion years after the photons travel, again, we see the wave function collapse. Our observation a billion years later has influenced the photons a billion years in the past.

At this point, this is about as far as we can explain quantum entanglement qualitatively. The next topic will be into how to use these principles in a quantum telescope and then onto other areas.

Quantum Telescope

As we consider a modification to the previous example of a satellite observing entangled photons, let us consider that one satellite is well staged above the north pole in an orbit. A second one is also well staged in orbit above the south pole, however, just a hair of distance to the rear of where incoming photons will be received. This time, the star emits one photon from the top of the star and another at the same time on the bottom of the star. The Earth-bound satellites are designed such that only very particular angles of the light are accepted. Since we don't have enough incoming light coming in to determine that there is in fact a star there, we are looking into a spot of darkness and very slightly modifying these angles that they are accepting, just as we might be triangulating outwards more and more until we find the star we cannot see with light because the incoming beams are too infrequent and can easily be coming from any sort of background *noise*. Now, let us discuss the photon path from this far away star. Based on the very far away star and light waves coming from the top and bottom, there is a similar path long enough that the pair of photons entangle, and after a long time, diverge. They are at such an angle that one hits the telescope on the north pole. This dictates the second photon immediately take an opposite spin as the other. This recording is perfectly timed. This second one, now having the opposite spin established travels that hair distance more, having its spin and registers opposite than the one at the north pole. This is then recorded.

Keeping both satellites focusing and recording incoming beams, we know that noise (photons that have nothing to do with what we are trying to observe) will also hit our satellites to try and throw us off. Now, let us say we collect a million data points (or more for that matter), we can see that as a function of time, maybe only 1% of those recordings show a perfect correlation of exactly matching opposite pairs and the 99% incoming do not have a confidence in opposites. The 99% can be filtered as noise. The 1% demonstrate a high confidence of entangled pairs. Based on the subtle angle between the two, we can triangulate those pairs to indicate the location of a star which could not be discovered by photons giving us an actual visual photo.

Slightly modifying the angle back and forth and observing these opposite pairs, not only can we triangulate the location, but we can also triangulate its diameter. For a normal satellite, it would require a larger mirror to see a faint star further away. For this quantum telescope, as long as the precision of measuring entangled photons in terms of timing and opposite spins, the same satellites can both just get more distance between them (like a higher orbit) and that would allow for a longer range of triangulation. For even further triangulation, again the same satellites can go at opposite ends of the solar system, so long as that same precision in measurement is calculated.

Quantum Communication

Quantum communication uses basic principles of quantum entanglement for the purpose of communication at long distances instantaneously. There are serious challenges to the idea, though some of the seemingly impossible barriers are being worked on with some success. While we are still far from its full potential, it is still worth discussing.

Einstein led to this "spooky action at a distance." He was also one of the greatest critics of his own work. The basic idea is that two particles might have say their own spin for example. Maybe we put them close enough together for a small amount of time and they become entangled. At the point of entanglement, both of the particles no longer have spin. They will remain without this quality until one of the particles is measured. Upon being measured, the one might have a positive spin and at the same moment, no matter how far away, the second particle must be of the opposite spin, negative. The fact that one particle might remain local at Earth and the other maybe on Mars, Einstein took issue with the fact that somehow information is now traveling faster than the speed of light, which he established as the galactic speed limit.

As an analogy, which is far easier said than done, let us store many sets of entangled particles within two boxes. Half of the pairs of entangled particles will ship to Mars and half remain in a box here at Earth. Having many pairs, the easy part is just taking a handful at a time and measuring them at distinct times that would me measuring for a set quick time and not measuring at other times. Based on the time of measuring and not measuring, this represents a digital (rather than analog) signal that is translated into information.

That is the easy part. We have the means to entangle particles. The hard part that currently prevents all of this, and I do not have solutions for. The biggest challenge we have to store these in a way that they remain untangled "in the box" until ready for use.

One theory that allows us to communicate instantaneously is that with the entanglement, the particles are transported non-locally via a spaceship to Mars. The thought process would seem to imply that since the non-local box had to be transported at subluminal speed that this bypasses the idea that there was ever FTL in the first place.

Another challenge, other than getting this box of entangled particles elsewhere is the fact that once both boxes are in place, we would have to have some sort of sensor on the non-local box to sense when some of these measurements are occurring non-locally. If we managed to keep them in a loop with say a magnetic field, we would have to sense when some of these are leaving such a thing.

With the many challenges that face quantum communication, it is appearing more and more that we are probably a long way from achieving it.

Quantum Teleportation

Quantum teleportation (think Star Trek transporter) is much like quantum communication but a far more complex problem. There needs to be a method of completely taking the 3-D image on at least an atomic particle's size, if not smaller. That image needs to be communicated to a quantum printer via quantum communication all at once. The hope here is that this process is not just getting a perfect replica of a body, but instead the momentum of all of those particles in the right direction as well. The size of a quantum computer processing such a thing is a far away process even if we had quantum computing working today. Even still, if it were possible, it is not necessary that with the momentum solved for each cell that we are not still just printing what would be a dead body and no soul.

I do not foresee this being a process used for organic bodies even if possible. It is far more likely that we would figure how to create miniature wormholes instead. It doesn't involve being taken apart and would likely be far more feasible as well. The additional benefit for a wormhole, once we understand the mechanics, we would not have to bring the non-local box in the first place. The idea of connecting two time-spaces via a wormhole seems to be agreed upon that its mechanics would not violate the laws of FTL, though that is what is seems to be doing.

Mitochondria

A part of the cell called the mitochondria is considered the power house of the cell. It is quite different than other parts of cells. It appears to have a lot of similarities to bacteria. It is thought that long ago when cells were first growing on Earth that mitochondria may have arrived via a comet. From there, the first mitochondria was able to enter a living cell at the time and became one of the most important symbiosis that has since existed.

From a physics point of view, it also has one quality that may have yet to be studied in detail. It has an ability to act as a proton pump for the cell, and more importantly, the ability to make these protons get very close to one another. What I question is whether or not this process actually makes these protons close enough for long enough to make the protons in a quantum entangled state and to what degree that has an effect on physiology. I might be more surprised if a bio-physicist had not already studied this. I simply haven't heard of such a study yet though.

Very often, much of our applications in science come from natural biological processes found in nature called biomimicry. As we look at new species of plants found in the Amazon, we find that something like a plant poison intended to ward off insects may have a numbing or sleepy feeling induced in humans. Rather than just finding this out and growing this new plant for pharmaceuticals is often a barrier, being not so economical. Instead, pharmaceuticals will attempt to isolate the specific compound and mimic its creation in a laboratory.

It should be of importance to note that most species of plants and their properties have yet to be discovered. As this is very often our starting point in medicine, this exemplifies the importance of keeping these undiscovered plants from extinction. At least here in the US, another important element due to our means of drugs to be allowed through the FDA, we often use a method called off-branding. What this means is that when a drug is approved for a specific purpose, the effort that went into the study was extensive and expensive. If a drug that is originally approved for a factor such as mania, but it also has a side effect of causing someone to get drowsy and sleep, the drug can still be issued as an off-brand for sleep even though it was never intended for that purpose. Why spend the extra money on another study when there is no incentive and it is still allowed to be prescribed as an off-brand. It is simply not in the financial interest of any party to do so.

This off-branding appears to be more tribal knowledge among doctors and not established best methods among doctors.

As a physicist, I have found that when I use the term *fundamental research*, most non-physicists hear fundamental and equate it to simple. That couldn't be further from the truth. If I would relate it to mind, soul, and body, I would say that putting a band-aid on a cut is knowledge. When you get to the one or two things in the mind and soul that are the really deep facts that make the soul want to continue on, that is fundamental. It is really the deep core that makes things tic. To that end, understanding atoms and molecules are important, it is the study of quarks that seem to go no further that is the fundamental research.

In either case, one other biological practice that I find of interest is that a few years ago, there was much study of fetal stem-cells that was driving science. It was not without controversy though. It has since been studied by bio-physicists and it appears that we can now take different types of cells and reprogram them from say a bone marrow cell and reprogram it to turn into a stem cell. From there, we have reprogrammed it to be any other kind of cell, whether a blood cell or a skin cell. The only limitation I have heard with the current ability to do this is we cannot yet do this to scale. Bio-physics is a relatively new field. Imagine being able to take a cell from a person and regenerate it into new heart muscle tissues and use an organic 3-D printer to make a replacement heart for a heart transplant, but unlike a donor heart, this is made of your own DNA and a person would not need to take a daily medication for the rest of their life for the potential that the body might reject the heart transplant. It is a fascinating field that I see only growing exponentially in the future.

Venus

Venus is a fascinating planet. Most recently, there has been a biomarker suggesting from the atmosphere that there in fact might be life present. What I find most fascinating isn't even potential life, because it might only be an anerobic cell that is creating this biomarker, but with the many similarities it has with Earth, it also has a huge energy supply already on the planet that would make it one of the easiest to terraform.

Venus is somewhere around 900-degree Fahrenheit. If we look at what makes a nuclear reactor work, there is a primary coolant that runs in a loop. It has a turbine that is turned as the nuclear reactor provides the heat. This is not much different than the solar collector as well, using concentrated light beams to provide the power instead of radiation. The same could be done on Venus, but the heat source is in the air on the surface of the planet. It would seem for a long time that this source of heat energy is endless. This energy could even be used to cool the inside of metal box with the refrigeration cycle running on some of this gained energy.

If we take this a step further and this metal box being able to travel and do some basic mining from the surface, we could also consider this metal box having the ability to sort and process raw materials and even create replicas of itself. After a few hundred years of this replication and these mobile mining boxes have an additional duty of separating carbon dioxide, the planet would cool as the green house gasses are processed into simpler components. As we study the planet more, it might be possible to find large areas that would be better for mining than other areas and possibly more specialized robots could be sent that focus on creating mineshafts that make necessary raw materials available to the surface for the other type of self-replicating robots. Fun fact, mining for lead will not likely be done for a long time as I hear it is liquid at the surface of Mars.

The Birth and Composition of Neutron Stars and Blackholes

The evolution of a star to a blackhole has become fairly well understood. A star like our Sun will primarily burn its hydrogen via fusion and produce helium. This fusion process can continue until it gets up to its heaviest element of iron. It is around that point that gravity is overpowering strong force that keeps the atomic structure from collapsing in on itself.

From this collapse, the mass of a large star collapses into a diameter of just several miles. Its spin is also relativistic. The very outer shell of this neutron star is composed of normal materials. With the processes within the neutron stars, it is expected that some strange types of quarks can exist.

The neutron star has a strong gravitational pull, many times larger than a regular star. As conditions change, such as sucking in enough matter and getting bigger, the neutron star will also collapse forming a blackhole. During this explosion, conditions are ripe for creating materials heavier than iron. One such explosion occurred recently where it is believed that part of the mass ejected from the explosion was a chunk of gold with the mass of the Earth. Space mining for the gold is complicated that it is traveling at relativistic speeds.

Rare encounters can occur. Knowing where and when become the biggest challenge. For example, if a blackhole gets close enough to a neutron star, the gravity of the blackhole wins and begins to tear the neutron star apart and absorbing all of its mass. In such an instance, if we had the when's and what's, we should be able to observe the inner components of a neutron star during this destruction. The next question might be what frequencies that we want to measure this occurrence. Inferred might only show us the changing heat. Visible light may not give us a clear picture of the exotic quarks. I believe an astrophysicist is going to be clearer in articulating.

Collision of blackholes do occur offering us potential of peering inside, however, I believe the rate that would occur would be pretty fast.

One observation that might shed some light into these potential collisions a bit more is that we have found galaxies that crash through one another. If such collisions are to be found, it would make sense that this would be a solid starting point to possibly find these collisions over other points in space.

The Edge of the Universe

The James Webb telescope is our most powerful yet. This allows us to look much further back in history than the Hubble telescope. As we consider the big bang, it is possible that the edge of the known universe is still not only traveling at close to the speed of light, but the universe is also expanding. To that end, what might be seen today is going to speed up with the expansion, making more and more of the seeable universe get to the speed of light with respect to Earth. This would mean that as time goes on, we will continually loose our ability to see outwards.

If this outer edge is continually getting faster and contracting all around us, there is a possibility that this accounts for some of the dark matter and dark energy. To what extent, since we cannot see it may very well be beyond our technology, or at least something that we have to observe a hundred years of change to put a better finger on how much.

Someone asked me a hypothetical question where rather than the moon is traveling within a hollow sphere yet the sphere had the same amount of mass as the Earth. The question was if the moon would still orbit within this structure if the sphere lied outside of the moon's orbit. I initially considered a sphere with an electric charge. Within a hollow sphere, a charge collects on the outside of the sphere so a charged particle on the inside would not move. On the other hand, if we consider a massive body starting from the surface and moving to the inner surface of the hollow sphere, this is different, say at the effective north pole, we could call the $+\hat{z}$ direction, at any point moving upwards, the $\pm\hat{x}$ and $\pm\hat{y}$ vectors of gravity would negate each other. The influencing gravities remaining are the $\pm \hat{z}$ gravities.

At the center, these two negate each other. On the other hand, as moving towards the north pole, the resultant gravity in the $-\hat{z}$ would be building, meaning that at the end of the line where the body reaches the inner side of the sphere on the north pole, the sum of the resultant would be pulling the body at the full gravitational acceleration equal to gravity on Earth, so the body would be inclined to again fall back into the center. Even if we used that inner spot of the north pole as a beginning point, assuming no air resistance, the body would be in free fall until reaching the center and after the center, it would begin slowing down until the point where when it reached the south pole, it would decelerate into a perfect standstill.

If we apply the same principle to the expanding universe such that at the edge while more and more mass all around us are seeing less and less because it is moving away from us at the speed of light relative to us, we would not see what is causing gravity to pull in our visible universe, but even more so, this increasing gravity is traveling at us at the speed of light. This would fit the bill for dark energy and mass. It is demonstrating that gravity would be pulling galaxies away from each other and it would be invisible. As many theories have tried to offer an explanation for dark energy and dark matter, I do not claim to have an answer, but it is one answer that seemingly fits the bill.

The Economy of Space Mining

Mining in space has some fundamental differences compared to that of Earth. Each project, whether it be on Mars or from comets and asteroids poses their own unique economies.

An asteroid called 16 Psyche contains enough gold to make every person on the planet a billionaire. If that were brought back and distributed on Earth, it would completely dilute the economy of gold on Earth and turn out to be rather useless endeavor in the end.

The current economy of spaceflight has its own uniqueness to it already though. For every pound of cargo brought up into space, regardless of the material, whether it is gold or water, it costs more than a thousand dollars per pound to transport it into space. Therefore, if we come up with a process to extract water from the moon at a cheaper price in the long run, this is going to be more economical than taking water to space. As we consider other metals that can be mined, such as copper, iron, or whatever, all of these materials are inherently worth more, not because of what the material is, but because of where the material is. If a long-term mining of copper can be done at even half of the cost of transporting to space, this is a huge advantage economically.

With several private space companies well established over the last decade, it is safe to say that more will develop and space is going to be one of the biggest booms for a long time coming. If someone tried to tell me that several space companies would be developed when I was a kid, I would not have believed it. It takes the pocket book of the US government to keep NASA afloat. Now, NASA is relying on some of these companies to keep the international space station afloat. More intriguing is that each of them are creating their own differences that set them apart. For example, Space X is heavily invested in reusing rockets. Virgin Galactic seems invested in space tourism. The creation of space hotels is already seriously being considered. It is not a far cry that we are not that far from having long-term ships being built. It might also be that massive spaceships (possibly generational) will be built for the purpose of leaving the solar system with no desire to return to Earth. In such a case, the bulk of the building materials will not be processed from Earth at all. It would be far more economical to mine asteroids and transport those materials to a shipyard in space. Thus, where materials are located make them more valuable than what the materials are, including the ice of comets.

The economy of mining Mars poses its own uniqueness. There is far less gravity on the planet which makes it easier to transport materials off of the surface. The economics still exist that if you wanted to export materials off of the planet, there is still the current issue that fuel for such a transport must be shipped from Earth currently. This single fact might make transporting materials not so economical. However, if these materials were to be going to the orbit of Earth for use in building ships in a shipyard, it might be economical after all.

Mars has another really unique feature that makes it attractive for mining. If we compare the surface are of Mars to that of land on Earth, they are very comparable in size. We currently do not do mining for materials (except oil) in the oceans. In compositions of minerals and metals, it is very likely that the amount of materials mined on Earth is very likely similar in amount as that found within the surface of Mars. Considering all of the gold ever taken from the land of Earth, there is likely a similar amount to be taken from Mars. The same process of how Earth collected those materials over millions or billions of years is very likely very much the same as that on Mars. Until the bars of gold are processed and put into a depository, it is anyone's game. Having such a depository that remains on Mars may be enough to give it a value. It would get that value from the idea that it could hypothetically be brought back to Earth if it was really wanted. Its just as true for copper or platinum. I think as we get to a point of a serious colony on Mars, it might remain more valuable on the surface of Mars in a depository rather than transported back to Earth though.

Over the next few decades, one of the greatest benefits we will see on Earth due to space exploration, in my opinion is because of the absolute necessity to recycle even the most insignificant things, we will adapt many of those recycling technologies to our future way of life. In California, there are complaints of water shortages. They stand at a distinct advantage to that in space. In California, they could with relative ease desalinate the ocean water next to them. In space, the only option is to completely recycle all water.

Western US Drought

Droughts occur all over the world. In the western United States, many of the reasons are probably better understood than any of the others in the world today. Despite this understanding, not nearly enough is being addressed to handle the situation. While it may be unfortunate that agriculture is the biggest player causing the most damage here, everything that is happening in California (along with other states) is everyone else's business too and that might have a silver lining. This means that solving the crisis is not just important to California, but all shareholders across the United States as we all require their goods when they are not able to grow during the winter in the east.

The easiest observation in what the problem with agriculture in the west is a lack of infrastructure. If your local grocery store was having issues, whether with refrigeration units or even just economic issues, the company owner would not just open shop next door in an empty field without walls and a ceiling. There is an expectation that a grocery store has a roof and walls. Expectations drive other technology over just trying to save some money, such as front doors that open and close on their own. Grocery stores do extensive research into where to put each of their goods and how to best market their products inside the store as well as external marketing.

The key here with all businesses, with grocery stores being just one, is adaptation. When issues arrive, the best solution is rarely to cut back. The better solution is always research problems, invest smartly, adapt research in small batches, and scale the problem. Often this method is used to increase efficiency and profits, but not always. Sometimes with the way you are doing business, you have already got the efficiency that you are going to get. In these instances that you have done all you can do with this business model, you have to radically change the business model itself and start over in the efficiency process.

I am going to begin with a flagrant violator of water abuse, the almond tree. It currently requires 1.1 gallons of water for a single almond nut. There are other factors that are important to consider when changing how almonds are grown. Trees live 20-25 years. The first 3-4 years, there is no yield of fruit. Trees grow from 13-40 feet high. What we see is there is not going to be a single silver bullet for everything, but instead, every type of crop is going to have their own properties that will drive some differences. Where almonds are heavily dependent on pollination, many crops are self-pollination, changing the dynamics.

All stakeholders need to invest smartly into a new way of doing business. This responsibility is not just farmers either. Grocery stores are a large stakeholder as well. If trees are diseased or a drought is causing less almonds to be available for sale, grocery stores have to change prices and consumers pay a price. To this end, consumers are stakeholders as well. Consumers need to be informed of the difference between responsible farming practices and understand what they are buying, just as they can now do in the grocery store between organic and non-organic products.

Here is my proposition, considering constraints as mentioned above, such as pollination and heavy water requirements. Almond trees require a genetic diversity in pollination so consider a single greenhouse that holds ten full size trees and accommodations for an additional 2 juvenile trees. This setup has an identical setup right next to it with a beehive as a connecting junction between the two. To overcome the largest problem with growing almonds, the enclosure is completely sealed from the ground to the top. It has an additional feature that allows for any rainwater that hits the roof to enter the greenhouse, but does not release any. Within the greenhouse is also a composting bin.

I recommended having accommodations for an additional two juvenile trees (which is more of an arbitrary value until researched for a better number). The goal here is to have each of the greenhouses to have a fruiting tree available on hand for when a tree is about to die. The math is fairly simple here, if a thousand trees are on hand, divide that by 25 years and the farmer can reasonably expect that he will have to cut down and replace 40 trees per year. I will point out more than once about the special ability of trees to prune and bonsai until wanting to allow it to grow to its full potential more than once.

The output product of an almond taken out of the greenhouse has very little water coming out of it. As a tree leaves the tree, going into the compost pile, virtually all water remains recycled in the greenhouse. As it evaporates and hits the ceiling, the water condensates and then drips back down onto the tree again. It is likely a reservoir for excess water ought to be kept and is able to receive and send water to other greenhouses as needed. The process of harvesting almonds requires a tree shaker machine to come in to release the almonds so a door to access the trees is necessary.

Almond trees are completely reliant on bees for pollination. They are so much dependent that massive amounts of bee hives are brought to the orchard and rented out to the farmer. In this greenhouse, a network of tunnels of sorts sufficient for the bees ought to not only allow for bees to do their job inhouse for the greenhouses, but also have a way in and out of their own from the greenhouse. An additional byproduct from caring for these bees is a specialized honey, *almond honey*. Specialized honey seems to carry a premium on costs.

If you ask a farmer how many trees he had, I believe his answer would be in the number of acres. With this new greenhouse system, he would be able to answer down to the tree and its level of maturity. To that end, this compartmental method of greenhousing isolates diseased trees as they appear, far limiting their ability to spread disease. Once seen, a dozen trees can be culled rather than culling or treating hundreds of trees.

Researching collectively is where within a specific subset of farming, in this case for any type of almond farming, all farmers in this endeavor as well as grocery stores take an investment into trying small changes in a small portion of each greenhouse. The goal is to create one standard model that seems to work best. Modifications might be in soil changes or different amounts of compost introduced. More importantly might be lighting capable of adding UV rays to the stock and finding out how much the UV rays added in different quantities improve tree quality, longevity, and product quantity increases. One further consideration with UV lights is that the lighting ought to be adjustable in height. This avoids spreading losses for shorter trees. To the same end, being able to test and adjust temperatures is another research element that might allow for more harvests per year.

It might also prove mathematically to take the proactive step of cutting a tree down early, say every 18 years rather than waiting until a tree has died. This is where I mentioned adapt research in small batches. Labor might be saved in the proactive of having to prune more towards the end of the cycle. Minimizing the needed height of the greenhouse is a factor. I will point out here that as data is gathered and centralized, minimizing is not tribal knowledge, but instead modeling an equation and taking a math derivative. This is not arbitrary. The same occurs from modifying UV rays up and down and again, going back to a math model that demonstrates what is best for the tree, maximizing output.

Forward thinking and invest smartly is essential. For a farmer that owns a thousand acres of trees, I would not recommend to switch to greenhouses completely within a year or two. However, I will point out that once this infrastructure is built, you own it and will not have to buy it over and over again. It will require basic maintenance depending on the elements included, such as heating / cooling or lights. Another reason I suggest slow growth is for the same reason that I see in how datacenters evolve. As we have suggested that for several years, I am proposing different variables in temperature, watering, etc…, it might be that the research shows a structure a few feet larger of a greenhouse. It would be reasonable to replace a few or build on to a few but unreasonable to reconstruct a thousand. Because of the number of variables that likely have not been completely resolved, it would be easier for all of the farmers to modify a few greenhouses rather than a single one to have to replace out a full farm's worth of greenhouses.

Farmers are smart people and usually very resilient. Growing a crop is very much a science. For nutrients the almond tree takes the most out of the ground, they are likely to know that growing say an onion that is perineal in the surrounding dirt takes in little nutrients and replaces it with the nutrients the tree needs, needing for little fertilizer to add. The onions in this case may not be the right addition. It might be a dandelion. For every different type of crop, this added crop will be different. The maximizing factor for this additional plant is not to be harvested, but to replace certain nutrients. Of course, perineal are the better choice, even if it does not provide 100% of the needed nutrients because of the inherent calculation that no additional work would have to be added in replanting them.

With this research, we are getting to a point where we are now getting computer scientists and mathematicians involved. Having thousands of farmers providing thousands of greenhouse data coming in, it is not needed to hire thousands of mathematicians and computer scientists. Instead just a few of each can modify and crunch the data provided by all of them. It is a small investment for a large payout.

I mentioned the size of a tree ranging from 13-40 feet tall. It is unnecessary to build every greenhouse tree to 40 foot tall just because some can get that high. It is likely that for the tallest tree species that a nice rate of return tends to maximize around 25-30 feet (where ever that researched value comes in) and the few trees that are approaching the 25 to 30-foot range are simply pruned.

Scaling the problem as better data comes in, particularly in the beginning strong research is essential to scaling greenhouses as a permanent solution. As suggested, I recommend that farmers not completely adapt to greenhouses right away. Further, as research maximizes a height for a tree, it would be advantageous for initial sets of greenhouses to also have the ability to scale vertically. By this I mean that if the trees are initially set to grow to 25 feet, but a taller structure of 30 feet is more advantageous, these initial greenhouses ought to be able to be improved on and have the ease of adjusting that height.

Most of the work in this book is geared towards future science and I would be amiss if I didn't go beyond just the next few years. So, here are my thoughts into the future tree farming that will heavily pour into farming in space and on Mars, starting here on Earth.

On a mission on the way to Mars and on Mars itself, it is not only impractical for astronauts and martians to depend on space food transported for the trip and habitation of Mars, it is just a plain bad idea. It would be a better idea to add the seat and accommodations for a space and martian farmer. Making room for a 30-foot tree in either case would also be impractical in either scenario. As we can currently do, an I do believe this is the next evolutionary step in tree farming, robotics and bonsais will dominate the field of farming. Particularly when heating / cooling and maximum lighting requirements for trees are well established, the same artificial intelligence AI principles will maximize fruit produced for a bonsai version of a fruit tree.

[Figure 1]

Take this example of a bonsai apple tree. By the way, bonsai is more indicative of the size being small, but literally meaning it was planted in a container. In particular, note the resources around the fruit rather than the fully-grown apple itself. Pruning of the roots, branches, and leaves has maximized the growing of a single apple. Proper fertilization is essential on a more regular basis because of the roots are quite small. Watering properly and more frequently is required because of the nature of a bonsai. Pruning is essential to maintain the size. With the help of AI, further maximization into even what amount of surface area of leaves for a single fruit is something that can be determined.

Of further importance is the notion that if an apple tree is grown from seed, its fruit will have a far different tasting fruit than the parent. In practice, most are not the taste that people like. It also requires several years to grow a tree before it produces its first fruits. Space flight and Mars farming are not going to have the luxuries of researching or testing as we might have here farming on Earth. So, rather than growing seeds, a cutting is made from a branch with a bud on it and soaked in water for a time. This cutting will not only grow roots and begin a new tree, it will also have an identical DNA and fruit of the parent. Particular to the apple tree, orchards grow an intermittent crab tree for pollination requirements. The tree does not require as many years before fruiting as well.

Almond trees have a bit of difference in that there is a larger requirement for genetic diversity in pollination. Therefore, rather than having all similar DNA trees and the occasional crab apple tree, the almond tree requires that ten or so genetically different be staged strategically (and tracked as the space farm is increased) such that this requirement can be fulfilled. An AI could more easily track the diversity requirements as it takes cuttings from trees and ensure the proper diversity for this specific type of tree is maintained.

Even with a fulltime space farmer on board, for all of the different crops to be tended to would probably be overwhelming. Therefore, it would be advantageous to have AI and robotics doing the work of daily observing the farming operations and status of every tree, pruning the different parts of the tree to minimize all resources such as branches, height, fertilizer requirements, etc... such that the single fruit is maximized while resources are adequate, but minimized.

The primary goal is that on the trip to Mars, there is a surplus for the space crew, producing fruit more on demand, but also so that cuttings that are closer to just being ready to fruit are ready for transport to Mars farming operations. Having robotics well trained in the art of bonsai is going to be essential to all elements of the space flight itself and the long-term habitation of Mars. Because of the ration of resources to end product, it is inevitable that this technology will also heavily spill over into farming worldwide. As we consider drought worldwide, such farming will likely be the only means of successfully growing adequate quantities of product. Using the small amounts of water available very carefully may be the only means of growing crops at all.

From the war in Ukraine, we have seen the effects that globalization can have. While often globalization can have a positive effect of offering a huge variety of products to sell, the basic staple of wheat and the product of fertilizer world wide have taken a huge hit worldwide. To this end, if we instead had a more decentralized system in some cases, the reliance of key areas having the power to be a weapon would be significantly reduced. For example, if this bonsai farm that also grows vertical, the amount of resources would be far less, making it easy for any other country to fill that very small gap needed rather than an entire large surface area that would need to be fertilized.

Other Considerations for the Spaceflight to Mars

I would suggest there is a difference in inhabiting Mars rather than living on Mars. Especially being a submariner myself, I have seen the effects that a long deployment of three months in close quarters has taken on people. I have a unique advantage in understanding some of the factors that come into play in such environments. Among most any occupation that will drive successful spaceflights to Mars, sailors that have sailed submarines for several years will make spaceflight far more successful.

One quality that will have to be addressed prior to flight is closterphobia and this is not as easily said than done. I have two examples of personal experience that highlights this. First, all submariners are required to visit aboard one of the boats on the waterfront while in submarine school. In those school days, I boarded the USS Seawolf with a fellow sailor. It was the roomiest boat on the waterfront. Submarine service is completely volunteer for mainly this reason. The fellow sailor quickly became very uneasy and realized that he could not do the tight spaces like he thought he could. That same day, he began to process out of submarine school and to the surface fleet where there was far more room.

While on my first deployment, a fellow sailor snapped after two weeks into the deployment. I will spare the details though he was quickly restricted to quarters and removed from the boat while at sea as soon as he was reasonably able to.

Even after the best of testing, there is no way of accurately projecting closterphobia perfectly. What exasperates the problem of travel to Mars is not only the longer journey to Mars, but critically, knowing that if there is a problem, there is no means of changing the mission or backing out in any way.

Another important fact with the psychology and a reason that we out to build several spaceships or more, roughly equal distant (besides being able to assist in a ship malfunction) is because if a person is realizing that the trip is something they cannot complete for any number of reasons, just knowing that there are break points such that a person could cancel the rest of the journey significantly changes how a person is just going from one point to the next, knowing they could back out. This actually makes it much easier to fully complete the journey. A person is just making one step to the next. To this end, with several ships in route to and from Mars, it might be advantageous to go ahead and plan on docking with another ship anyway.

Another reason it might be advantageous in counting on this docking might be for resupplying. If we consider the mass of fuel used for example, if one ship is making a journey carrying all fuel and other supplies, they are using extra fuel to carry it to Mars and all the way back home. On the other hand, let us assume there are several ships traveling this path, then that extra fuel that the ship returning home is not carrying the fuel for say the last $\frac{2}{3}rd$'s journey. This means that the fuel needed to get $1\frac{1}{3}rd$ of the journey is not having to be carried and using fuel to use by the first spaceship. This might even be the selling point for using a couple of spaceports. The journey would not be as quick, but it seems the fuel savings for a while would be advantageous. Not only for fuel, but the same would go for any types of supplies to be replenished, this would be helpful.

If we are going to take the bold trip (particularly if we want a long-term presence and a serious colony), to Mars, we must be ready to account for all that can go wrong plus the problems that can occur with the human element. For the first, right about every issue that can occur on a submarine, there was always a backup system. Even while I was driving, there were two gauges in front of me for depth for example. Each had its own piping to the outer hull and independent gauge so that if one went offline, the secondary I could still count on. Even for the mission itself, being ready to deploy nuclear weapons anywhere in the world, if my boat was destroyed, we had another boat that was covering the target package. Redundancy was always a factor in everything that we did in the submarine force.

This redundancy is going to have to be a critical component of Mars travels. Here is what will not work. The James Webb Space Telescope at launch until opening up at its target point had over 300 points such that if any one of them went wrong, the telescope would fail. Submarining is a dangerous business. Even modern submarines over the last twenty years have had boats and crew lost to sea. I had a couple close calls myself. For every issue that might cause a catastrophic failure, redundancies or multiple redundancies must be put into place. Such redundancies must be good enough that if it is something that must be done manually, the primary system failure must allow time for a person to switch to redundancy. If this condition of time is not possible, then an extra mechanism must be in place such that an automated system has adequate time to adjust to redundancy.

I've watched a documentary about two rovers being sent to Mars. The hope was that one was to survive. I can almost buy this as there were no people aboard. The rovers encountered an intense solar flare and the computers on the rovers had to be wiped and reinstalled. What really concerns me is that solar flares are known to be unpredictable, but also a known phenomenon. There is no good reason that we would not include adequate shielding for this phenomenon for every mission sent to space.

Further, even for circuits and wiring, we currently have well established mechanisms to these devices. Here are a few that should be addressed. The easiest is a shield of a material that will properly insulate from the solar flare. Second, just as we know how to protect wiring from an electro-magnetic pulse EMP by shielding it in a metal mesh, all wiring can be done the same. Third, almost all wiring on a rover can be eliminated, thus no wiring could be affected. This would be done by primarily using fiber optics. While the rovers are in flight, power cords would be disconnected from batteries. A separate mechanism aboard the craft transporting the rovers would instigate the power cords to connect to the battery closer to the last moments of flight to Mars.

I made the case for using multiple space stations along the path of Mars, being points of replenishing supplies and fuel. If we had these stations which are well stocked for years' worth of operations at any given time, let us say we do not want to slow down and dock with the station as this increases fuel consumption. Here is an adjustment to the process. While the station is relatively stationary between Mars and Earth, as the spaceship is going past, a small shuttle carrying the fuel and supplies for this one specific spaceship traveling can launch a bit early and match course and speed of the spaceship and dock with the spaceship. Rather than the spaceship having to dock, a significantly smaller shuttle is making a minimum trip for the supplies and fuel. From there, the small shuttle would return to the station.

I think to sum all of this up, I would suggest that slow is fast. It is essential to take the time to consider psychology, redundant systems, and also a plan B when there are catastrophic events.

The Earth does not Orbit the Sun

We have all been lied to, but not a significant amount. It turns out that we do not orbit the Sun, but rather the Earth is orbiting where the Sun was eight minutes ago. Even this, while more accurate remains incorrect. It all comes down to the simple equation of gravity and the speed of gravity. For gravity:

$$F_g = \frac{Gm_1m_2}{r^2}$$

So, for the Sun and Earth, the force of gravity F_g remains accurate, but because there is a delay, it is traveling where the Sun was eight minutes ago. However, all of the planets, comets, asteroids and such are all playing a tug-a-war with all of the other bodies, but also on that delayed schedule.

If the Sun went out 8 minutes, gone is a new orbit in 8 minutes. To further see how out of sync this is, Jupiter is 43 light away from the Sun while traveling at 29,000 mph and Pluto is about 5 $\frac{1}{2}$ light hours away from the Sun while traveling over 10,000mph.

Trying to sum all of these different forces together to get a single moment would be a really difficult thing to calculate by hand. Luckily, our computers have the ability to assist us.

Depending on whether we are closer of further from Mars, there is a significant delay here as well. If we did some basic math for a rocket and simply sent the rocket for where we currently seen it visually, the rocket would completely miss its target.

I mentioned some oddities in the section of *The Gravity as the Oddball* of some issues with the galaxy. Knowing now that the

Earth does not orbit the Sun, but where it was, we begin to have a real problem with the Milky Way. The Milky turns at 1.3 million miles per hour, or 2.1kph. With respect to outside our galaxy, the Milky Way is moving at over a million miles per hour. Or solar system, that is fairly close to the center of the galaxy at only 27,000 lightyears. Still, as our instruments are well tuned enough that we can tell how all of the orbiting is done within our solar system, we are exacerbating the mechanics when looking at ourselves in the galaxy. Our sun is orbiting a supermassive hole moving over a million miles an hour, but on a time delay of 27,000 years lag time. Further out, towards the end of the Milky Way is working on a lag time between 50,000 to 100,000 years ago. That trajectory doesn't seem reasonable to keep the galaxy cohesive.

A separate small point that many people know already, but needs to come out somewhere, so here it is. As we look at stars in the sky, we are not looking at the star as it is today. However, many lightyears away the star is, that is determines the age of the picture we are seeing. For example, if we look at a galaxy with our fancy telescope and get an image one billion lightyear s away, that means that it took a billion years to travel from that galaxy to our planet. That image is of it was a billion years ago. Further, if we find an exoplanet that is 1000 lightyears away today and it had highly intelligent life, maybe beyond our own technology, that picture is a thousand years old. If they just found our planet today as well, they would be looking at us in Medieval times before the invention of electricity. That other 1000 years that we are not seeing of the exoplanet may have inhabitants that have since destroyed themselves with nuclear weapons or died off though a plague. We would have no way of knowing other than waiting another thousand years to wait and see. We would still have that lag time of another 1000 years though.

Nuclear Physics

Nuclear physics is a fun field. I would be doing a disservice if I did not hit at least a couple of ideas about it. When basic chemistry is explained, it is often looking at a nucleus of the atom and giving the shapes of how the electrons create these three-dimensional volumes and shells. I would suggest that while this is one way of modeling what is going on, viewing them in terms of equations is another that offers some convenience at times.

We currently have four fundamental forces. Strong forces, weak forces, and electromagnetic forces are three that we can combine with each other to make sense of the world. The weakest is gravity and we have not been able to mesh that with the other three yet.

We know from electromagnetic forces that like charges repel one another. It makes sense then that all of the electrons are going to avoid one another so these shells make sense in that this shows us that electron shells do not allow other electrons to hit one another. What it does not answer is how the nucleus of the atom allows the protons to be so close to one another.

This is where nuclear physics kicks in. Beginning with weak force, this force only works on really small distances. When protons begin to get closer, on a scale of 10^{-18}m, weak force causes the protons to move away from one another. At a distance of 3×10^{-17}m, the weak force exponentially decreases and is 10,000 times weaker. At a distance of 10^{-15}m, strong force that holds the nucleus together. It is reasonable to consider that the protons are vibrating between these distances and this causes them to hold their positions, even more so than the electromagnetic force acting on them.

If we look at the periodic table, we also see that for stable atoms, on the most part, atoms are most stable when there is roughly a similarity in the neutrons and protons. Considering any one of these atoms that are stable, nuclear physics has mapped out a single type of atom, say carbon, and looked at their stability in more neutrons or less neutrons. As the number of neutrons diverge from the number of protons, we find that the average half-life significantly decreases. Qualitatively, we can say that even though the neutrons are not acting in an electromagnetic capacity, it is mediating a stability of the protons in the nucleus. So, it does seem to serve a purpose.

I have questioned for a while now, is it possible to use a positron within an atom rather than a proton. The neutron and proton are roughly 2000 times more massive than an electron. If it were possible, in particular, substituting a neutron for something like an antineutron, but also having 2000 times lighter mass, then we could construct a new material, like stainless steel, but 2000 times lighter. Not only would this float in the air more than a hydrogen or helium balloon, but it would have the strength of steel. This would certainly allow us to create a space elevator. My initial hunch is that while that might be cool, it does not seem feasible. Substituting a positron for a proton would not likely be controlled by strong force and weak force in the same way. It is also no guarantee that neutrons would offer a stabilizing force for the positron nucleus either.

And lastly, here is what nuclear physics tells us about nuclear warfare. Uranium in a nuclear reactor is controlled by firing neutrons into the atom. In a nuclear reactor, this is a process that is done slowly in order to control how much heat is going to be generated in the loop of water. The opposite is true for a nuclear weapon. Many neutrons are thrown at it for a quick reaction. Every time a neutron hits the nucleus, the uranium atom splits. In this splitting process, four of the neutrons from the split uranium are expelled. These four go out and find other uranium atoms, causing an exponential decay, rapidly growing until there is enough energy released to cause the mushroom cloud we are all familiar with.

There is an added layer of conventional explosives that are around the uranium warhead though. It has a precise distance from the uranium and perfectly timed such that when the warhead is about to explode, the conventional explosives explode just a fraction of a second after. This extra explosion causes the mess of splitting uranium atoms to hold its position for just another fraction of a second, allowing this rapidly growing free neutrons to keep hitting other uranium atoms more. For just that fraction of a second, it is enough to get an exponential number of neutrons doing their job, making the warhead exponentially more energetic. Theoretically, it could be possible to offer one more layer of explosives to do the same and still get more bang for the buck. Since it is breaking more of the uranium atoms down, this might even make for a relatively cleaner explosion in that not as many uranium atoms are going to be distributed among the devastated landscape. Don't get me wrong, the spent materials are still nothing to be happy about.

Friction

In high school and even in beginning chemistry at the university, friction has been explained to me as two surfaces such that when we look under a microscope, even for a polished surface, we would find microscopic edges that cause friction between the two. This may be further refined in advanced chemistry that I have not seen; however, I believe this model to be seriously flawed. My view is very different than what I have been taught. A quick point that gets me to this is considering two sets of two different materials. The first set is highly polished and the second set is smooth, though not highly polished. If we test these two sets for their friction, we get the same results and the microscopic differences are not accounting for any differences.

Just to have the equations and definitions, we have two types of equations of friction, static and kinetic. Static friction is that friction that you initially have to get over before you actually begin moving something. Kinetic is the friction you encounter after it begins moving. In terms of labeling them as forces, we have:

$$F_s = \mu_s N$$

$$F_k = \mu_k N$$

N in this equation represents the normal force. If on a level platform, it is simply equal to mg, or its weight. For μ_s and μ_k, these are experimentally found values. It is as simple as looking up two chemicals and these two numbers pop up, though a google search is probably more mainstream today than a book.

This is another reason to doubt the notion of microscopic differences. There are not separate values for a polished surface and a smooth surface. They are the same.

I am going to make a large separation between a friction and an effective friction. Friction is as I have mentioned above. Effective friction would be like tape being sticky or wheels with spikes traveling on a road. While these effective frictions have real world applications, such as improving tires on a snowy road, this can create a model for the application, but is not friction at all.

I am proposing that friction is an interaction of two materials that has to do with valence shells interacting between atoms, and for molecules I would say is similar to atoms, but more complicated as the path that electrons take in molecules are also more complicated.

If either model can explain how friction is working in a system, why would it be important to focus on a distinction? Let us consider that we are engineering a new tire composition for a car. For most applications, the simple process of looking up the values would be sufficient. In this case of a tire, we can use known friction for rubber on the road to do many things. However, adding new chemicals to a road would be enhanced by understanding the why's of friction. It would be easier to make a list of materials to integrate into the new tire with specific properties.

Having much experimental data available for friction, I believe the next step is looking towards how valence shells of atoms are working with one another. The more complicated version of this is going to be how molecules work together. For example, we can slide solid sodium across a surface like iron and get a value for friction. But, when we use salt (NaCl) and slide that across iron, we find a completely different result because the electrons are being shared between the atoms of salt. I believe these shells interacting is what causes these differences in friction.

The Sound of the Sun

Most people don't really think about it, but the Sun is constantly releasing energy equivalent to many nuclear bombs firing all of the time. We see the Sun far away, but we rarely think of the Sun as having sound. On the Sun, there is plenty of sound. This is also true for other planets, neutron stars, and for a blackhole just outside of the event horizon. Sound requires a medium to travel, unlike a light wave. Small idea but interesting none the less.

The Not So Standard of Mass

In 1884, replicas of the standard for the kilogram were sent worldwide so that other countries would have an accurate example so they could also have the standard of the kilogram within their country [1]. After some time, many of those replicas were again compared to the original. Ironically, some of those replicas either lost or gained mass.

One oddity that comes up is quantum fluctuations. Even in the vacuum of space, the vacuum is quite energetic. Every once in a while, a set of particles and antiparticles come into existence and the combine, leaving existence as if it was never there. Even more rarely, sometimes one of these particles do in fact escape. To not violate the Law of Conservation of Mass and Energy, the first sets that end up negated resolve themselves. It is assumed that if a particle like an electron does manage to escape, somewhere else the average of its antiparticle also escapes, also giving a net of zero between the two.

As a consequence of this quantum fluctuation comes the Casimir effect. If we have two metal plates with a separation of 10nm in a vacuum of space which is about 100 times the typical size of an atom, this should simply separate with no resistance, but instead this effect produces a force of about 1 atmosphere of pressure. [2]

Bypassing all of the complicated physics equations, this Casimir effect has demonstrated the ability of having an energy density to be negative with respect to ordinary matter vacuum energy. Many physicists believe that this may allow for a stable traversable wormhole. Even if we could create a really small one, it should be possible to create one that could send and receive information.

Applications of Science

If we tried to put our finger on one thing that has allowed the technology around us to exist in such a way that it had never exploded in the past, what would that be?

Looking at ancient Egypt, we can see that astronomers of the time were very dedicated to their field, given the tools available at the time. So, it had nothing to do with dedication.

What about the standardization of interworking parts just a couple hundred years ago? While there were obvious benefits, that isn't quite it either.

And, maybe it was the collection of knowledge for the sake of finding value in knowledge? The Great Library of Alexandria was established between 285-246BC. Estimates very that somewhere around 100,000 books worth of material lasting for several hundred years. So, that is probably not it either.

All of these, including the creation of the scientific process helped, but I would suggest that instead, it is more of a small group of people began to look at a phenomenon and then ask the question, what can I do with this? Having knowledge and looking at how it can be used is a large divide.

Legend says that Sir Faraday demonstrated to the king a battery with two wires and the wires would repel or attract. The king asked "Wonderful, but how good is this?" Faraday's answer was "I can't tell you now but one day you can tax it." He of course couldn't be more right.

 Application of science is the art of having a discovered or known phenomenon and applying that to a current problem. Sometimes these are big leaps and sometimes they are small. The invention of the transistor was a major leap and took the place of short-lasting vacuum tubes. A small leap from there was slight modifications to acquire the technology of a Light Emitting Diode (an LED light). The slight modification in materials happened to give off light. As the transistor was the backbone into the microprocessor, putting many LED's together now provides efficient lighting that is long lasting.

Besides looking at current technology, we can look at small things today that seem to have no application and begin to change the line of questioning. We don't need to have a major breakthrough today, but we can start with small concepts that over time will seem to evolve into bigger and bigger principles later. For example, biologists have long understood that the mitochondria work. We then change the question to if this is acting as a proton pump, can we demonstrate the ability of this to perform quantum entanglement applications? What principles of the well-established solar collector can we use for propulsion of spacecraft? Even the small observable facts can lead to great technology in the future. Edison could not have imagined his contribution to modern computers and artificial intelligence today.

Credits:

[Figure 1] - Apple tree bonsai photo: backyardboss.net

Page 73

[1] https://www.wired.com/2013/01/keeping-kilogram-constant/

Page 95

[2] https://en.wikipedia.org/wiki/Casimir_effect

Page 95

[3] https://energyeducation.ca/encyclopedia/Nuclear_fusion_in_the_Sun

www.ingramcontent.com/pod-product-compliance
Lightning Source LLC
Chambersburg PA
CBHW072143170526
45158CB00004BA/1493